Wireless & Electrical Cyclopedia

Catalog

No. 20

The Electro Importing Co.

233 FULTON ST.

NEW YORK CITY

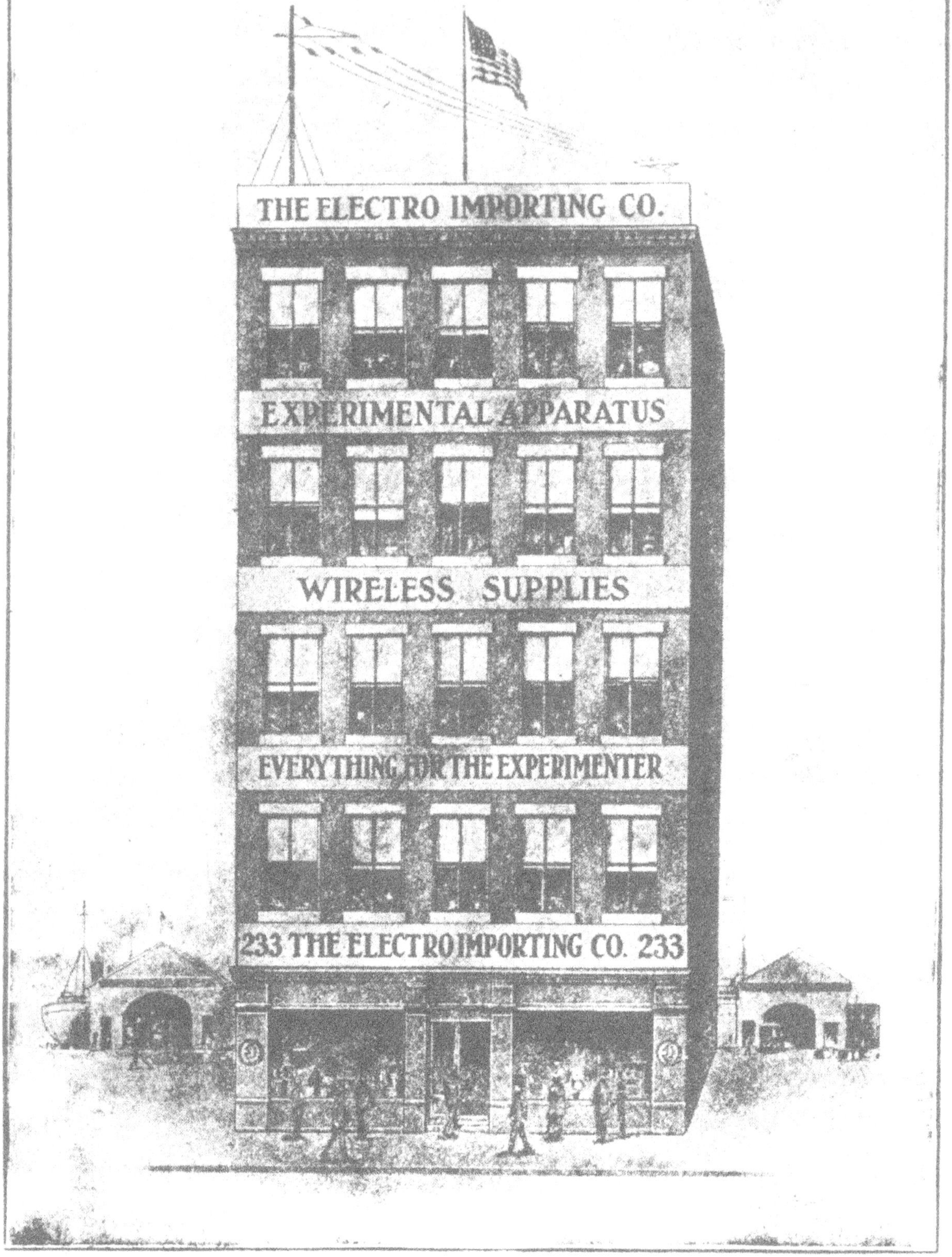

THE ELECTRO IMPORTING CO.

EXPERIMENTAL APPARATUS

WIRELESS SUPPLIES

EVERYTHING FOR THE EXPERIMENTER

233 THE ELECTRO IMPORTING CO. 233

CATALOG No. 20

First Edition 1918
300,000 COPIES

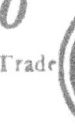

Trade Mark

The Electro Importing Co.
233 Fulton Street
New York City

NO ORDER FOR LESS THAN 50 CENTS ACCEPTED

BEFORE ORDERING PLEASE READ CAREFULLY.
IMPORTANT
All Previous Prices Are Withdrawn With This Issue

OUR GUARANTEE:—We guarantee every article listed in this Catalogue to be equal in every detail to the illustration and printed description, but as we are constantly improving and bettering our goods there may be slight changes from details as shown in cuts. We will replace free of charge any article or part thereof, in which there may be a mechanical defect of construction if same is returned prepaid to us within five days after receipt.

CABLE AND TELEGRAPH ADDRESS "ELIMPORT," NEW YORK.
TELEPHONES 7777 and 7776 Cortland.

OUR TERMS:—Cash with order. We do not open accounts with private individuals, nor do we sell on the installment plan. OUR LOW PRICES DO NOT JUSTIFY IT.

Always allow sufficient money to cover postage. Weight of packages can be readily figured from the weights given in the catalogue descriptions. Knowing the weight and the parcel post zone in which your post office is located, measuring from New York, you can easily figure the amount of postage required from the parcel post rate table shown on page 5.

EXPRESS C. O. D. ORDERS for at least $5.00 or more will be accepted by us East of the Mississippi. WE REQUIRE A DEPOSIT of 25 per cent. of the amount purchased.

FREIGHT C. O. D. ORDERS for $3.00 or more will be accepted if 25 per cent. of the amount is remitted with the order.

REMITTANCES should be made by N. Y. Draft, Post Office or Express Money Order by Registered Letter. Do not send money unregistered by regular mail. Out-of-town checks accepted only if 10 cents exchange is added.

U. S. STAMPS (new and in good condition) will be accepted instead of cash in amounts up to $3.00. Above this amount 5 per cent. to cover brokerage fee must be added to total.

SHIPPING DIRECTIONS should accompany each order; in their absence we will use our best judgment in making selection of routes.

WHEN ORDERING give catalogue number.

RETURN OF GOODS:—Occasionally an article does not work as you think it should, or it may become defective through mishandling in transportation. In that event don't send the goods back without first writing us a letter stating just what is the trouble. Frequently we can advise you of a remedy by mail which obviates the wait necessarily incident to returning goods. If we can't advise you how to remedy the trouble we will tell you how to return the goods. Never return goods without having marked on it your name and address and in the package a slip of paper with your name, address and order number. This insures the maximum of speed possible in correcting an error or trouble. Always write why you are sending goods back for we can't guess it, though it may be obvious to you. Goods returned without our permission are returned at customer's risk. We do not accept goods which are sent express or freight collect.

GOODS BY MAIL AND PARCEL POST. We are not responsible for goods lost or broken in the mails. For your own protection, order mail goods insured.

Fragile articles will be carefully packed and duly labeled by us, but as the Parcel Post does not guarantee their safe delivery we cannot be held responsible for breakage or lost shipments. For your own protection, order Parcel Post goods INSURED. The fee for this insurance for each package is:

$0.03 for $10 Insurance $0.05 for $25 Insurance $0.10 for $50 Insurance

WE ANSWER CHEERFULLY and promptly all inquires for special prices and discounts on quantity orders. Don't accept propositions until you get our prices.

WE CARRY A COMPLETE STOCK of all the listed goods in our New York house, and in most cases ship the same day that order is received.

OUR ENGINEERING STAFF will cheerfully answer, free of charge, any and all technical questions pertaining to our goods, if a 3-cent stamp to cover postage is enclosed. if diagrams for connections are desired an additional charge of 10 cents will be made by our drafting department. WIRELESS QUESTIONS, "Hook ups," etc., not bearing direct relation to our goods are charged for at the rate of 10 cents each. Where special calculations or special information is desired, we will inform correspondent as to the cost of such work.

NEVER USE THE ORDER BLANK for communications, questions. etc. It will surely delay the answer if you do. Write on a separate sheet, which can be transferred quickly to the right Department.

TO WHOM IT MAY CONCERN.

WHEREAS, the Electro Importing Company is a manufacturing corporation, incorporated to do business in the State of New York, and

WHEREAS, Hugo Gernsback is the President thereof, now therefore, I, Hugo Gernsback of New York City, New York, being duly sworn depose and say that:

WHEREAS, each and every testimonial published at the foot of each page of this catalog is unsolicited, that these testimonials have been received from bona fida customers, that their names and addresses are genuine as published, that the testimonials are unaltered and undoctored, and,

WHEREAS, the testimonials as published have been taken at random from the testimonial files of the Electro Importing Company, and that they represent only a small percentage of the total number of these letters in the possession of the Company, now therefore:

It is agreed that the Electro Importing Company will pay the sum of $100.00 to anyone who will prove that the above facts are not true and correct as affirmed.

Hugo Gernsback

Sworn to before me this 29 Day of October 1912

E. E. Junior

Notary Public Kings County

Certificate filed N.Y. County #2

E. E. JUNIOR NOTARY PUBLIC KINGS COUNTY

The E. I. Co. owns and controls 14 patents. They guarantee you valuable and exclusive features and construction.

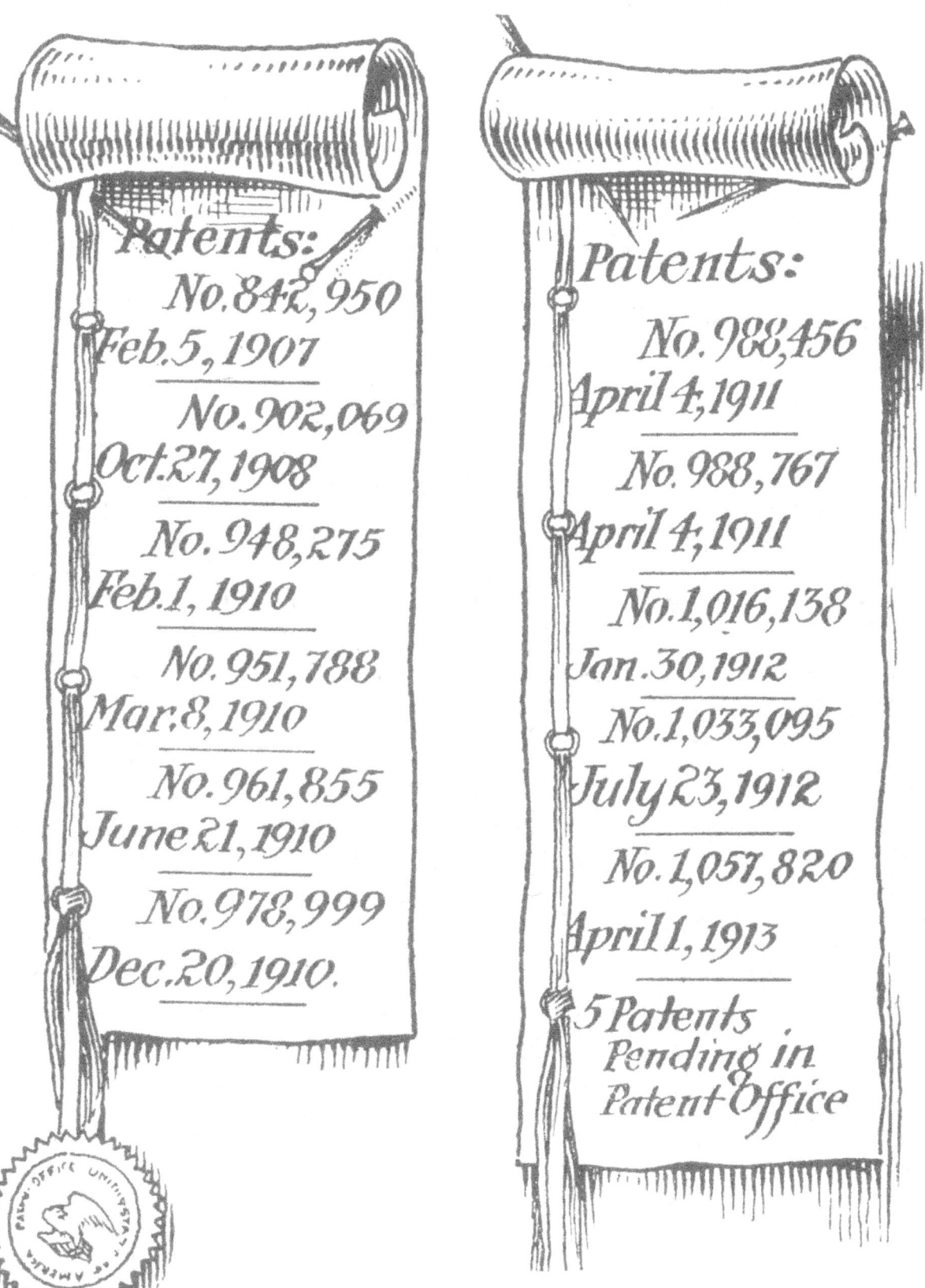

Patents:
No. 842,950
Feb. 5, 1907

No. 902,069
Oct. 27, 1908

No. 948,275
Feb. 1, 1910

No. 951,788
Mar. 8, 1910

No. 961,855
June 21, 1910

No. 978,999
Dec. 20, 1910.

Patents:
No. 988,456
April 4, 1911

No. 988,767
April 4, 1911

No. 1,016,138
Jan. 30, 1912

No. 1,033,095
July 23, 1912

No. 1,057,820
April 1, 1913

5 Patents Pending in Patent Office

IF YOU DON'T KNOW US THIS WILL TELL YOU

The Electro Importing Co. was born in 1904 at No. 32 Park Place, New York, in a little office 10 feet square. At that time we were the only and original concern in America selling solely experimental electrical goods.

The first amateur wireless outfit in America was made and sold by us, and the writer, the pioneer in Amateur Wireless, is not without good reasons called: **"The Father of Amateur Wireless"** by his many friends and followers.

We grew rapidly—not because we had the goods to sell, but because we gave everybody a square deal, and in 1905 we moved to larger quarters at 87 Warren Street, where we had better manufacturing facilities.

Early in 1908 we again had outgrown our facilities and moved to 80-82-84-86 West Broadway, with six times as much floor space as before.

These quarters in less than one year proved inadequate, and in 1910, when conditions became intolerable, when every inch of floor, wall and even ceiling space was at a premium, we looked around for a new home and moved to our own 5-story building at No. 233 Fulton Street, which we now occupy from basement to roof.

Although we have here over 15,000 square feet at our disposal, we again became cramped in 1912, and we had to take the building next door to us, thus almost doubling our floor space.

It must be plain to anyone that there must be a good reason for such a growth. There is no secret about it. The answer is: **ENORMOUS VALUE FOR THE MONEY, AND THE FAMOUS E. I. CO. SQUARE DEAL.**

This means that your dollar will go twice as far when spent with us than with any other electrical house in the U. S. Also, as everyone knows, the **E. I. CO. ALWAYS MAKES GOOD.** That's the key of our success.

Our mail last year averaged 1,500 pieces of mail daily. We receive and ship from 400-600 orders each day. We control more patents on experimental and Wireless apparatus than any other concern in America. We manufacture and handle more such material than any other of our competitors.

We buy our raw material by the ton and carload, where our competitors buy in pound lots. Do you wonder that our prices are from 25 to 50 per cent. lower, and that we give you superior goods on top of it?

Our trademark: **"Everything for the Experimenter"** is not an idle phrase. Ninety per cent. of our customers are experimenters, and our international reputation as **"THE electrical mail order house"** is well founded. There is not a civilized country on the globe where our instruments are not known.

We realize more than anybody else that the average experimenter's pocketbook is not overfed, and our prices are accordingly. We could easily get twice the amount for some of the articles, but our policy has always been to make a very small profit on a large amount of goods, thereby invariably giving the customer the benefit.

In closing let us add a word to the wise. If you pay more for goods, such as we list, **YOU ARE BEING ROBBED.** If you pay less, you will unfailingly **PAY MORE,** for you will positively get inferior material, which you must replace later.

WE DO NOT SELL GOLD BRICKS, WE CANNOT AFFORD IT.

THE ELECTRO IMPORTING CO.

H. GERNSBACK, President.

HOW TO FIGURE PARCEL POST RATES

With every article in our catalogue, we state the exact shipping weight of same.

Having this weight, and knowing the zone in which your post office is located, measuring from New York, you will find the correct amount of postage applying to the goods you are ordering, in consulting the Parcel Post Table below.

U. S. Parcel Post

POSTAGE RATE TABLE FOR PARCEL POST SHIPMENTS

Zones	Local	1 & 2	3	4	5	6	7	8
Weight		The Parcel Post Rate to Pay Is:						
1 lb	$0.05	$0.05	$0.06	$0.07	$0.08	$0.09	$0.11	$0.12
2	.06	.06	.08	.11	.14	.17	.21	.24
3	.06	.07	.10	.15	.20	.25	.31	.36
4	.07	.08	.12	.19	.26	.33	.41	.48
5	.07	.09	.14	.23	.32	.41	.51	.60
6	.08	.10	.16	.27	.38	.49	.61	.72
7	.08	.11	.18	.31	.44	.57	.71	.84
8	.09	.12	.20	.35	.50	.65	.81	.96
9	.09	.13	.22	.39	.56	.73	.91	1.08
10	.10	.14	.24	.43	.62	.81	1.01	1.20
11	.10	.15	.26	.47	.68	.89	1.11	1.32
12	.11	.16	.28	.51	.74	.97	1.21	1.44
13	.11	.17	.30	.55	.80	1.05	1.31	1.56
14	.12	.18	.32	.59	.86	1.13	1.41	1.68
15	.12	.19	.34	.63	.92	1.21	1.51	1.80
16	.13	.20	.36	.67	.98	1.29	1.61	1.92
17	.13	.21	.38	.71	1.04	1.37	1.71	2.04
18	.14	.22	.40	.75	1.10	1.45	1.81	2.16
19	.14	.23	.42	.79	1.16	1.53	1.91	2.28
20	.15	.24	.44	.83	1.22	1.61	2.01	2.40
21	.15	.25						
22	.16	.26						
23	.16	.27						
24	.17	.28						
25	.17	.29						
26	.18	.30						
27	.18	.31						
28	.19	.32						
29	.19	.33						
30	.20	.34						
31	.20	.35						
32	.21	.36						
33	.21	.37						
34	.22	.38						
35	.22	.39						
36	.23	.40						
37	.23	.41						
38	.24	.42						
39	.24	.43						
40	.25	.44						
41	.25	.45						
42	.26	.46						
43	.26	.47						
44	.27	.48						
45	.27	.49						
46	.28	.50						
47	.28	.51						
48	.29	.52						
49	.29	.53						
50	.30	.54						

Parcel Post Rates are computed according to weight of the parcel to be shipped and according to the distance between the shipping point and the delivery point. For this purpose the U. S. is divided into 8 Zones, with different rates of postage applicable to each.

The table shows the amount of postage by parcel post, according to the weight of the package and according to distance by zones.

If you don't know the Zone in which your post office is located measuring from New York, the Postmaster will tell you.

Parcels weighing 4 ounces or less are mailable at the rate of 1 cent for each ounce or fraction thereof, regardless of distance. Parcels weighing more than 4 ounces up to 16 ounces must be mailed as a full pound. If in figuring out the cost of your order, we discover that you made a mistake, and that your remittance was short we will forward the shipment by Parcel Post C. O. D. including in your shortage the U. S. Postal Fee. We trust that this manner of handling your orders will meet with your approval as it enables you to obtain your goods without long correspondence and the necessity of sending in the difference, so that the few additional cents for C. O. D. fees can hardly be considered.

Books are accepted at Parcel Post Rates. (One cent for each two ounces up to eight ounces, over eight ounces same rate as other Parcel Post Matter.)

Parcel Post
(Continued)

In some instances Express Rates are cheaper than Parcel Post Rates, consult the table:

Between NEW YORK and the following points:	5 lbs. Express Insured	10 lbs. Express Insured	20 lbs. Express Insured
Chicago, Ill.	$0.31	$0.42	$0.64
St. Louis, Mo.	.32	.44	.68
Denver, Col.	.47	.75	1.30
Butte, Mont.	.58	.96	1.72
Dallas, Tex.	.45	.70	1.20
San Francisco, Cal.	.71	1.22	2.24

The weights given in our catalogue are the exact shipping weights; this means the apparatus or article packed and boxed ready for shipment. In some instances, as with glassware, etc., it is easily understood that the wrapping must be heavy so as to insure the safe arrival of the goods. In other instances a paper wrapping only is needed with some excelsior, which amounts to fractions of an ounce.

If you send in an order calling for different items of our goods all to be packed and shipped together, in one package or box, it is understood that you will benefit a good deal on the transportation charges, as it will not take much more packing material for three or four small articles than for one. The saving which we can effect in all these cases is to your advantage as we will return to you in every case the over-payment of your transportation charges, if there is any.

Note.—Be very careful in figuring out the transportation charges for goods weighing, for instance, four ounces. If you order two items calling for four ounces each, this would make eight ounces shipping weight, but in reality you are required to pay for one pound, as the Parcel Post Law states that articles which exceed four ounces up to 16 ounces must be mailed as a full pound. The same holds true if one article weighs one pound and another four pounds and six ounces. This means that you would have to send in money enough to cover six pounds at the Parcel Post rate. Some articles in our catalogue, which are sold by the pound, are always to be figured PLUS the packing material. Thus, if you order one pound of tin foil it is necessary to send in postage enough to cover TWO pounds at the Parcel Post rate. If you order four pounds send in sufficient postage to cover FIVE pounds, and so on.

Parcel Post Service to Foreign Countries

Parcel Post Rates to: **PORTO RICO, CANAL ZONE** (Isthmus of Panama), **HAWAII, PHILIPPINE ISLANDS, TUTUILA** (Pago-Pago) and other parts of **SAMOA**, in possession of the U. S. and **GUAM** (Ladrone Island), are the same in all respects and conditions as **DOMESTIC RATES** for the "Eighth Zone."

Rates for **CANADA, CUBA** and **MEXICO** are 12 cents per lb. Limit of weight is 4 lbs. 6 oz.

Rates for Europe, including **GREAT BRITAIN** and **IRELAND**, most of the British possessions, also **NEWFOUNDLAND, AUSTRALIA**, etc., and **all other countries** to which the Parcel Post extends are 12 cents per lb. Limit of weight is 11 lbs.

How to Return Goods to Us by Parcel Post

When you return goods by parcel post, put the letter you write in an envelope and paste or tie the envelope securely to the outside of the package. In addition to the postage you put on the package, put a 3-cent stamp on the envelope.

Educational Institutions That Buy Our Goods

The following Educational Institutions have been supplied by us **REGULARLY** for years with our goods. The list which we give here is only a partial one; we have many hundred more institutions on our books, but lack of space prevents us from publishing same. We hardly need say that the names speak for themselves and no higher tribute to the quality of our goods and our excellent service which we give could be presented:—

Adrian Public School, Adrian, Mich.
Allegheny College, Meadville, Pa.
American International College, Springfield, Mass.
Aurora Public Schools, Aurora, Ill.
Board of Education, Dover Plain, N. J.
Baltimore Dept. of Education, Baltimore, Md.
Berlin School Dept., Berlin, N. H.
Billings Board of Education, Billings, Mont.
Bloomfield Theological Seminary, Bloomfield, N. J.
Board of Education, Huron, So. Dak.
Board of Education, Neodesha, Kans.
Brattleboro High School, Brattleboro, Vt.
Brookville Board of Education, Brookville, Pa.
Brown University, Providence, R. I.
Carnegie Institute of Technology, Pittsburgh, Pa.
Cherokee County High School, Columbus, Kans.
College of Emporia, Emporia, Kans.
Columbia University, New York City.
Colt Memorial High School, Bristol,
Connecticut Agricultural College, Storrs, Conn.
Cornell University, Ithaca, N. Y. R. I.
Dakota Wesleyan University, Mitchell, S. Dak.
Darlington Public School, Darlington, S. C.
Dartmouth Medical College, Hanover, N. H.
Delaware College, Newark, Del.
Department of Education, New York City.
Dubuque German College and Seminary, Dubuque, Iowa.
Elder High School, Cincinnati, Ohio.
Ferris Independent School District, Ferris, Tex.
Findlay College, Findlay, Ohio
Georgia School of Technology, Atlanta, Ga.
Gonzago University, Spokane, Wash.
Grove City College, Grove City, Pa.
Heidelberg University, Tiffin, Ohio
High School of Memphis, Memphis, Tenn.

Iowa State Teachers College, Cedar Falls, Iowa
Lake Placid High School, Lake Placid, N. Y.
Massachusetts Agricultural College, Amherst, Mass.
Michigan Agricultural College, E. Lansing, Mich.
Minden High School, Minden, Nebr.
Parson College, Fairfield, Iowa
Princeton University, Princeton, N. J.
Rock Hill College, Ellicott City, Md.
School District No. 2, Brattleboro, Vt.
St. Edward's College, Austin, Tex.
St. Joseph School, San Jose, Calif.
St. Mary College, Dayton, Ohio
St. Mary's Mission, O'Kanogan Co., Wash.
South Georgia College, Helena, Ga.
South Western Presbyterian University, Clarksville, Tenn.
State Manual Training Normal School, Pittsburg, Kan.
State University of Iowa, Iowa City, Iowa
Stevens High School, Clairmont, N. H.
St. Mary's Mission, Mission, Wash.
University of Fla., Gainesville, Fla.
University of Manitoba, Winnipeg, Canada
University of Nebraska, Lincoln, Nebr.
University of California, Berkeley, Calif.
University of New Mexico, Albuquerque, N. Mex.
University of Washington, Seattle, Wash.
Vassar College, Poughkeepsie, N. Y.
Yale University, New Haven, Conn.

Government Departments that buy our goods:

U. S. Navy Supply Dept.
U. S. Signals Corps, Field Co.
U. S. Bureau of Standards, Washington, D. C.
U. S. Coast Artillery School, Fort Monroe, Va.

TREATISE ON WIRELESS TELEGRAPHY

By H. GERNSBACK
President ELECTRO IMPORTING CO.
Editor "THE ELECTRICAL EXPERIMENTER"
Manager RADIO LEAGUE OF AMERICA

WIRELESS AND THE AMATEUR
A RETROSPECT
PART ONE

ON DECEMBER 13, 1912, the new wireless law went into effect. The average wireless "fiend" who has not followed the topic from the start will be interested in the following facts:

The very first talk about Wireless Legislation in the country started in 1908. The writer in his Editorial in the November, 1908, issue of **Modern Electrics** pointed out that a wireless law was sure to be passed in a very short while. In order to guard against unfair legislation as far as the wireless amateur was concerned the writer, in January, 1908, organized the "Wireless Association of America." This was done to bring all wireless amateurs together and to protest against unfair laws. Previous to this time there was no wireless club or association in the country. In January, 1913, there were over 230 clubs in existence, all of which owe their origin to the "Wireless Association of America."

The association had no sooner become a national body than the first wireless bill made its appearance. It was the famous Roberts Bill, put up by the since defunct wireless "trust." The writer single handedly, fought this bill, tooth and nail. He had representatives in Washington, and was the direct cause of having some 8,000 wireless amateurs send protesting letters and telegrams to their congressmen in Washington. The writer's Editorial which inspired the thousands of amateurs, appeared in the January, 1910, issue of **Modern Electrics**. **It was the only Editorial during this time that fought the Roberts Bill.** No other electrical periodical seemed to care a whoop whether the amateur should be muzzled or not. If the Roberts Bill had become a law there would be no wireless amateurs to-day.

That editorial quickly found its way into the press and hundreds of newspapers endorsed the writer's stand. During January, 1910, the **New York American**, the **New York Independent**, the **New York World**, the **New York Times**, the **Boston Transcript**, etc., all lauded and commended the writer's views. (See Editorial article February, 1910, **Modern Electrics**.) Public sentiment quickly turned against the Roberts Bill and it was dropped.

The first wireless bill not antagonistic to the amateur, The Burke Bill, appeared on March 8, 1910. It had some defects, however, and was dropped also. The Depew Wireless Bill appeared May 6, 1910, but did not meet with general approval; as the writer pointed out in his Editorial in the June, 1910, issue of **Modern Electrics**. It had several undesirable features, and the bill was never seriously considered, although it actually passed the Senate. (See Editorial, August, 1910, **Modern Electrics**.)

At last the Alexander Bill made its appearance on December 11, 1911. This bill as far as the amateur was concerned was not quite acceptable to the writer, who had the amateurs' rights at heart, and steps were immediately taken to bring about an amendment as the writer, perhaps more than anyone else, realized that this bill, in some form or other, would become a law sooner or later. This is clearly stated in his Editorial in the February, 1912, issue of **Modern Electrics. In that Editorial is to be found also the first and now historical recommendation that if a wireless law was to be framed it should restrict the amateur from using a higher power than 1 kw. and his wave length should be kept below 200 metres.** No one else had thought of it before, and it is to be noted that when Congress finally passed the present wireless law, it accepted the writer's recommendation in full, thus paying him the greatest compliment, while at the same time acknowledging the fact that he acted as the then sole spokesman for and in behalf of the wireless amateur.

In March, 1912, the writer, in a letter to the New York Times (See page 24, April, 1912, issue Modern Electrics) pointed out the shortcomings of the Alexander Bill, and protested against unfair legislation.

The Times, as well as a host of other newspapers, took up the cry and published broadcast the shortcomings of the Alexander Bill. All this agitation had the desired effect and Mr. Alexander for the first time realized that the amateur could not be muzzled, especially when there was such a periodical as Modern Electrics to champion his cause. Promptly in April the Alexander Wireless Bill, amended, appeared and here for the first time in history the amateur and his rights are introduced in any wireless bill.

Mr. Alexander and his advisers accepted the writer's recommendation as set forth in his Editorial in the February, 1912, issue of Modern Electrics. (See Paragraph 15, 2nd Part of this Treatise.)

It will be noted that it copied the writer's recommendations word for word. The amateur had at last come into his own. This is all the more remarkable as this is the only country that recognizes the wireless amateur. On May 7, 1912, the Alexander Bill, amended, now known as S-6412, passed the United States Senate and on May 8th was sent to the House of Representatives and referred to the Committee on the Merchant Marine and Fisheries. The bill was signed on August 13th by President Taft, thus making it a law.

This terminated the fight which the writer had waged single-handedly for almost five years in behalf of the American amateur. Now that it is all over, and that Uncle Sam has set his seal of approval upon the amateur's wireless, the writer cannot but extend his heartiest congratulations to the 400,000 American amateurs, and he furthermore wishes to extend his thanks to all the amateurs who have supported him in his fight to bring about a new wireless era in America.

Long live the Wireless! Long live the Amateur!!

WIRELESS AND THE LAYMAN
PART TWO

THE QUESTION we hear from most beginners is:—"What outfit do you advise me to use? I know nothing about wireless."

We advise the use of ANY of our receiving outfits. They are ALL good—the result of 12 years manufacturing. Which one to choose depends upon yourself, your taste and your pocketbook. This is where YOU must decide. Of course, ALL our outfits work, they are all guaranteed to do so— OR MONEY BACK. The lower priced outfits have naturally a short range—they won't catch messages hundreds of miles away, and those without tuning coils cannot be used to "cut out" one of the messages when two of them are in the "air" at the same time. It is self-evident,

> Receiving
> Wireless Messages

though, that you can start with the very cheapest outfit.—say anyone of our detectors and a pony telephone receiver. With such an outfit messages can be picked up astonishingly well indeed. Many of our enthusiastic young friends started with such an outfit and kept on adding instruments till they finally had up-to-date stations.

The next question hurled at us is:—"How can I receive messages if I don't know the codes?"

A wireless telegram, no matter if it is in Chinese or English, "comes in" in dots and dashes. When you have the telephone receivers to your ear and a message is coming in, you hear a series of long and short, clear, distinct buzzes. A long buzz is a dash, a short buzz is a dot. We sell a 10c. code chart by means of which the dots and dashes are translated into letters. Thus (in the Morse code) dash, dash, dot, stands for the letter G; dash, dash means M, dash, dot, dash, dot means J and so forth. Any person with a few weeks' practice "listening to the wireless" can master the code, and read the messages with ease.

Remember that there are over two thousand high powered wireless stations in this country alone, each being able to transmit messages of over a thousand miles distance.

There are almost at any minute, during night and day, messages in the air, no matter where you are,—sending YOU messages, only waiting to be picked up by you. It is truly wonderful; it is the cheapest as well as the most elevating diversion known to modern man, the most inspiring example of the triumph of mind over matter.

"How about the Wireless Law?" you want to know next.

The law does not apply for stations used for **receiving** only. **There is no law which forbids you to receive all the messages you wish.** You can receive as many and as long as you please,—Uncle Sam doesn't mind. But you MUST preserve the secrecy of the message. You must not make use of any information you receive by wireless, if this information is of such a nature that makes it private property. Your own conscience will tell you which message to keep secret and which one you can make use of. Here is the text of the Law:

SECRECY OF MESSAGES

"**Nineteenth. No person or persons engaged in or having knowledge of the operation of any station or stations, shall divulge or publish the contents of any messages transmitted or received by such station, except to the person or persons to whom the same may be directed, or their authorized agent, or to another station employed to forward such message to its destination, unless legally required to do so by the court of competent jurisdiction or other competent authority. Any person guilty of divulging or publishing any message, except as herein provided, shall, on conviction thereof, be punishable by a fine of not more than two hundred and fifty dollars or imprisonment for a period of not exceeding three months, or both fine and imprisonment in the discretion of the court.**"

Of late a great many stations are beginning to use the wireless telephone. This art is rapidly being perfected and is the coming thing in "Wireless." There is hardly a week that you do not read about some new wireless telephone and some new distance record established.

Wireless Telephony

It is of course understood that any receiving apparatus that can receive wireless telegraph messages, 90 times out of 100, can receive wireless telephone messages. Of course, in that case no code is required as the voice comes through the receiver the same as through the regular telephone. (For further details on Wireless Telephony, see Lesson No. 18 of The Wireless Course.)

The question asked mostly by the layman is: "How far can I receive with such and such an outfit, my aerial being so high and so long?"

Distance

Nobody can correctly answer such a question. You can reason it out as well as we can. For example: Would you ask us: "How far away can I hear the steam whistle of the X & Y Cotton Mill?" No, you wouldn't, for it all depends. First, how hard the whistle blows, second, how good your hearing is, third, how the wind blows, and fourth, how many and how great are the intervening objects between the whistle and your ear. Some days you may hear the whistle two miles off with the wind blowing your way. Or if you are way down in a cellar you may only hear it faintly, although you are but two blocks away from it. It all depends. The one thing you are sure of is that the whistle blows about the same strength each day. The same reasoning holds true for wireless to a very great extent.

As a rule, the higher up and the bigger your aerial, the better the wireless reception will be. Naturally if you are a thousand miles off from a station that can but send 500 miles, you won't hear it, no matter how good your instruments are. It's like trying to hear the sound of a whistle 10 miles away from you, that can at the very best be heard only within a radius of 5 miles. Just use a little horse sense and you can do your own deducting; no wireless expert is required. It is also evident that the messages cannot come in with the maximum loudness unless the instruments are well in tune, and unless well designed instruments are used. Thus a loose coupler will give louder signals than a small tuning coil. It also depends a lot on the detector and its adjustment.

This is the way the detectors range according to their sensitiveness:

1st, THE RADIOSON (the most sensitive detector to date); 2nd, The

RADIOCITE Detector; 3rd, The Crystaloi Detector; 4th, The Perikon; 5th, Zincite and Bornite; 6th, Silicon and Galena; 7th, Iron Pyrites (Ferron); 8th, Carborundum; 9th, Molybdenite. (See Lesson No. 10, of The Wireless Course, on Detectors.)

If you are entirely surrounded by high mountains or steel buildings, you naturally will not expect to receive messages as well as if you were on the top of a mountain. Also remember that wireless waves travel **twice as far over water as over land,** and that you can reach **twice as far after sundown than during the daytime.**

This seems to be the greatest stumbling block for most beginners. Again let us make a comparison. Take two pianos and place them in the same room. Or two violins will do as well. Tune two strings, one on each instrument, so both will give exactly the same note. Pick one of the strings in order to sound it, and the other "tuned" string, although 10 feet away will sound in unison, although you did not touch it. **Both are now in tune.** Both give out the same (sound) wave length. No mystery here. The secret lies in the fact that both strings ARE OF THE SAME LENGTH, and have the same tension, roughly speaking. Make one string longer than the other and both are "out of tune."

Wave Lengths and Tuning

The same in wireless. Nearly all commercial stations operate on a wave length of from 300 to 600 meters. (A meter measures 39.37 inches.) Now in order that you can hear such a station, you must be able to tune up to 600 meters; roughly speaking your aerial should be 600 meters long electrically. That, however, would b⌐ a pretty expensive and cumbersome aerial. Besides it isn't required. ⱱ ɇ simply wind, roughly speaking, 600 meters of wire on a coil or drum and our aerial can now be quite small, within certain limits of course, and we can for this reason "catch" the station having a 600 meter wave length, providing our other instruments are sensitive enough. By referring to our catalogue it will be seen that our No. CEK8486 tuning coil, as well as our No. DBE12002 loose coupler have sufficient wave length capacity to catch 700 meter waves. As they are both provided with adjusting sliders, more or less wire can be put into the circuit, and therefore both these instruments can be used to catch wave lengths from 100 up to 700 meters, but not over this amount.

Therefore, if we should want to hear a station having 1400 meters wave length, we would connect two No. CEK8486 tuning coils in series, which would give us 700 + 700 = 1400 meters wave length. Or we would connect one No. CEK8486 tuner in series with the primary of the No. DBE12002 loose coupler and we would get the same effective wave length. As a rule only stations doing long distance work use excessive wave lengths, thus the Marconi Transatlantic station at Glace Bay has a wave length of about 7100 meters, while the new Government station at Washington, which sends messages over 3,000 miles, has a wave length of about 2,500 meters. By consulting the **"Wireless Blue Book"** the wave length of all important stations can be found, as each station normally uses a certain prescribed wave length. (See Lessons No. 4, 5, 6, 7, 8, 9 of The Wireless Course.)

The best all around aerial is about 75 feet long, composed of four strands "Antenium" wire. One of the best forms is shown herewith. We recommend our No. AF10007 insulators, although others as listed in our catalogue can be used. For a 75 foot aerial, the strands should be about two to three feet apart. For a 150 foot aerial from three to four feet apart and so on. The strands should never be less than 1½ feet apart

Aerial and Ground

even for a very small aerial. All connections should be soldered if possible. Use as many insulators as feasible, remember you have but little energy when receiving; few and poor insulators waste 50 per cent. of the little incoming energy. If you have a good spacious roof it is not necessary to use poles to hold up the aerial. It may be stretched between two chimneys, etc. The spreaders to hold the wire strands apart may be of bamboo, wood, metal pipe, etc. If metal is used, the wire strands should be insulated from the former. (See Lesson No. 11, of The Wireless Course, on Aerials.)

The ground is quite important. The best wire to use is a No. 4 copper wire run from the instruments to the water or gas pipe using one of our No. AE10003 ground clamps to make an efficient connection. If no water or gas pipe is to be had, bury a metal plate, copper preferred, not less than three feet square, in a good moist ground; a number of these plates connected to the ground wire would be preferable. The heavy ground wire is soldered to the plate, of course. It should be buried at least six feet deep. Another good ground is a six to ten feet long iron pipe rammed into moist earth, the ground wire being connected to it, either soldered, screwed, etc. The ground wire running from ground to instruments should never be less than No. 16 B. & S. copper, and can, of course, be bare. Insulation on a ground wire is just that much waste.

Connections and Hook-Ups

The diagrams given in our catalogue show how to connect most of our instruments. Our Wireless Course (Lessons 12 and 13), give hundreds more of them, and our Engineering Department, on receipt of 10c. to cover postage, will be only too glad to furnish any hook-up to be used in connection with our instruments. Connections should be made with nothing finer than No. 18 B. & S. copper wire (Annunciator wire). All connections must be as short and straight as possible. Avoid all wire crossing as far as practicable; if you can't avoid crosses, the wires should cross each other at right angles; and NEVER wind the connecting wire in coils ("curls") which may look pretty, but kills all wireless messages. Make all connections as tight as possible, a loose connection is worse than no connection at all.

We presume you have a complete receiving set. You proceed thus:

Reception of Messages

First, you must know if your detector is adjusted to its greatest sensitiveness. If a message is just coming in, you will have little difficulty to adjust the detector to its best sensitiveness. If no message comes in you don't know if your detector is in its best receptive condition. (This does not hold true of the **RADIOSON** detector, which needs no adjusting.) For this reason, the up-to-date wireless man uses the "Buzzer test." Aside from giving imitation wireless buzzes, the buzzer set may be used to **practice telegraphy.** It consists of three things: 1st—Our famous No. HK1800 **RADIOTONE** (see illustration at left); 2nd—Our No. CE1118 key; 3rd—A dry cell. Connections MUST be made as shown. Now every time you press the key you will get a perfect imitation of a wireless signal and it becomes child's play to adjust the detector to its greatest efficiency. The buzzer test can of course be used with ANY detector. It saves lots of time and bother and is quite necessary. Sometimes a detector may have a "dead spot" and you might be "listening" in for hours, without being able to catch as much as one dot. The buzzer test makes such an occurrence impossible.

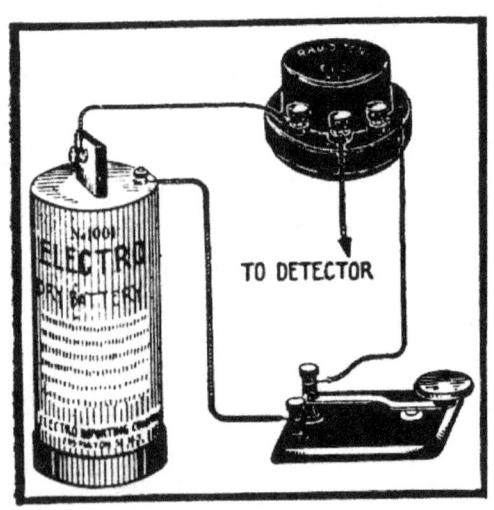

Only One Wire Goes to Detector

Of course to get the best results while testing your detector you should use our Radiotone. It is a test buzzer with its special connections all developed for just one purpose to test crystals. It is absolutely silent and can therefore be right on your instrument table.

13

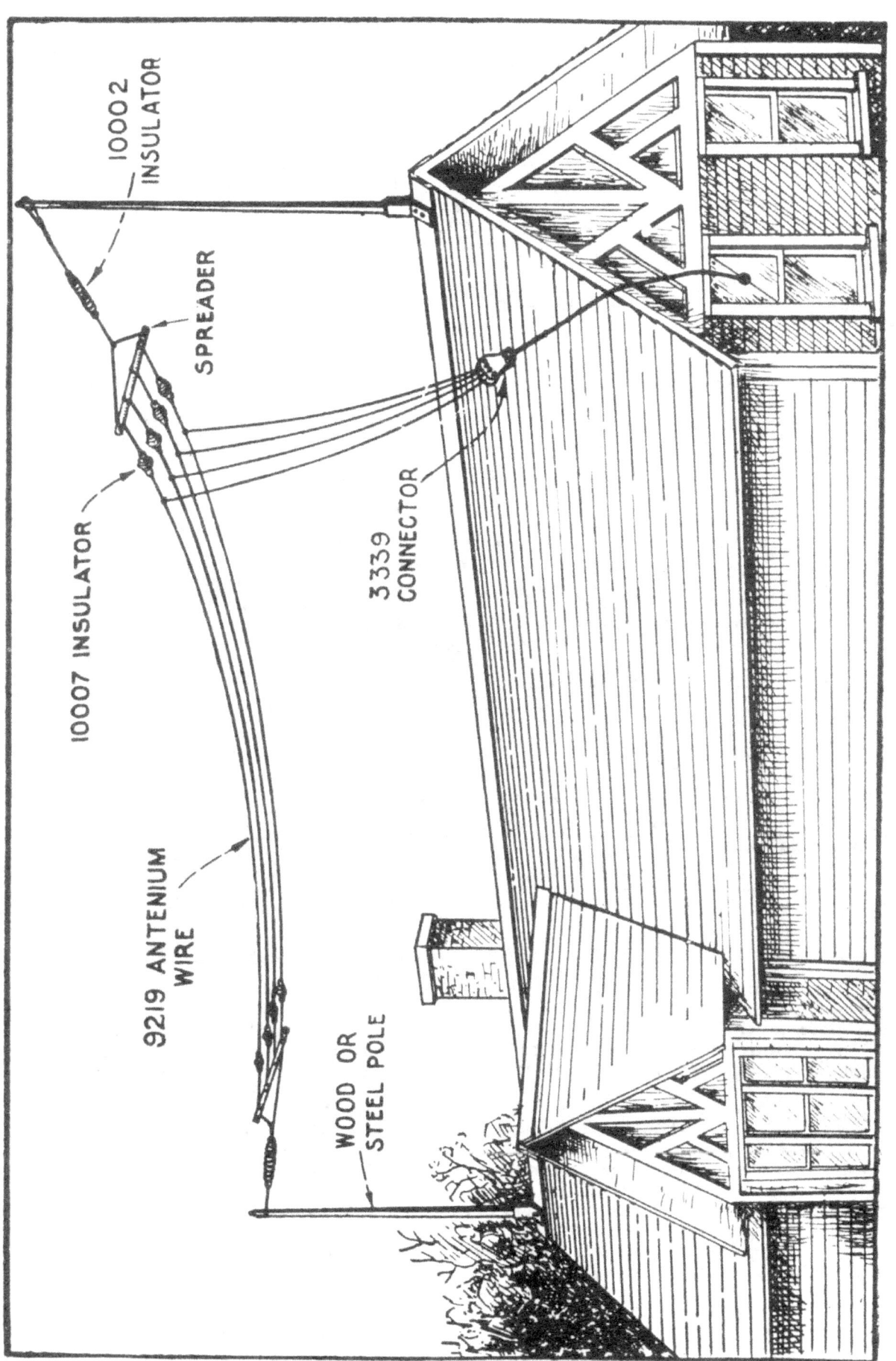

SIMPLE AERIAL CONSTRUCTION

10002 INSULATOR

SPREADER

10007 INSULATOR

9219 ANTENIUM WIRE

3339 CONNECTOR

WOOD OR STEEL POLE

When the detector is adjusted the tuning coil is regulated by moving the slider or sliders back and forth till the signals are heard the loudest. send messages to your heart's content, **and no one can tell you to stop** If the loose coupler is used the secondary is moved back and forth in addition, till the best position is reached. Now the variable condenser (or condensers) are adjusted if required.

The variable condenser is of the greatest use during excessive "static," which sometimes interferes seriously, during summer weather, especially when "taking" a long distance message. It is also of invaluable help to "cut out" unwanted messages when two or more are "coming in" simultaneously. Thus by adjusting the tuner (or loose coupler) in conjunction with the variable condenser it is often possible to cut out all interference from unwanted stations.

It is an excellent idea to have several detectors in a station, arranged in such a manner that by means of a multi-point switch any one of them can be thrown into the circuit. It will thus be found, that some stations, especially during interference, can be heard better on a certain detector than on another. Some will be found to work best for long distance work, others work best for medium distances, etc., etc., (see also "Wireless Course," Lessons No. 8 and 9).

Let us quote the law, as far as the amateur is concerned, before going any further:

THE WIRELESS ACT

| Transmitting Stations |

"**Be it enacted by the Senate and House of Representatives of the United States of America, in Congress assembled; That a person, company, or corporation** within the jurisdiction of the United States shall not use or operate any apparatus for radio communication* as a means of commercial intercourse among the several States, or with foreign nations, or upon any vessel of the United States engaged in interstate or foreign commerce, or for the transmission of radiograms or signals the effect of which extends beyond the jurisdiction of the State or Territory in which the same are made, or where interference would be caused thereby, with the receipt of messages or signals from beyond the jurisdiction of the said State or Territory, except under and in accordance with a license, revocable for cause, in that behalf granted by the Secretary of Commerce and Labor upon application therefor; but nothing in this Act shall be construed to apply to the transmission and exchange of radiograms; or signals between points situated in the same State; Provided, That the effect thereof shall not extend beyond the jurisdiction of the said State or interfere with the reception of radiograms or signals from beyond said jurisdiction."

GENERAL RESTRICTIONS ON PRIVATE STATIONS

"Fifteenth. No private or commercial station not engaged in the transaction of bona fide commercial business by radio communication or in experimentation in connection with the development and manufacture of radio apparatus for commercial purposes shall use a transmitting wave length exceeding two hundred meters or a transformer input exceeding one kilowatt except by special authority of the Secretary of Commerce and Labor contained in the license of the station; Provided, That the owner or operator of a station of the character mentioned in this regulation shall not be liable for a violation of the requirements of the third or fourth regulations to the penalties of one hundred dollars or twenty-five dollars, respectively, provided in this section unless the person maintaining or operating such station shall have been notified that the said transmitter has been found upon tests conducted by the Government, to be so adjusted as to violate the said third and fourth regulations, and opportunity has been given to said owner or operator to adjust said transmitter in conformity with said regulations.

*Wireless Telegraph or Telephone sending stations included.

SPECIAL RESTRICTIONS IN THE VICINITIES OF GOVERNMENT STATIONS

"**Sixteenth. No station of the character mentioned in regulation fifteenth situated within five nautical miles of a naval or military station shall use a transmitting wave length exceeding two hundred meters or a transformer input exceeding one-half kilowatt.**"

Let us explain in plain English just what this means: As you notice from the first paragraph, the part which we underlined, it is pointed out to you that the law does not concern you unless you send messages from one state into another. You therefore do not require a license as long as your messages do not reach over the border of your state and if you do not interfere with a station's business (in your state) which receives messages from another state. Of course, you want to know how you can tell what your transmitting range is. We will explain.

It has been proved by experience with spark coils, that in almost all cases a one-inch spark cannot possibly reach over eight miles. From this information the following table has resulted:

TRANSMITTING DISTANCES OF SPARK COILS

¼-in. coil, Maximum trans. dist., 2 Miles.

½-in. coil, Maximum trans. dist., 4 Miles.

1-in. coil, Maximum trans. dist., 8 Miles.

1½-in. coil, Maximum trans. dist., 12 Miles.

2-in. coil, Maximum trans. dist., 16 Miles.

3-in. coil, Maximum trans. dist., 24 Miles.

4-in. coil, Maximum trans. dist., 32 Miles.

And so forth.

With open core transformers the spark length cannot be taken as a basis to figure distances, but the input in watts is used. It has been found that for each five watts input into the primary, not more than one (1) mile can possibly be covered.

Thus our No. GGE8050 transformer coil in conjunction with our electrolytic interrupter uses 5 amperes at 110 volts. That is 550 Watts. Five goes into 550 just 110 times. The maximum distance that can be covered with the No. GGE8050 coil (unless you use more current) is 110 miles. If you use but one secondary the distance that can be covered is cut in half, or 55 miles. For closed core transformers we figure ten watts for each mile. Therefore, our No. BAEK9281, ¼ K.W. transformer will at best not send over 50 miles, and our No. DKX9282, 1 K.W. type will not send over 100 miles.

Knowing what the distance is, as the crow flies, from your locality to the nearest border of the next state, you can easily figure out what the maximum power is that you can use if you do not wish to take out a license. An example:

Suppose you live in the city of Columbus, Ohio. The nearest state line is Kentucky about 86 miles in a direct line from Columbus. If you do not wish to be licensed you can use any spark coil up to 10-inch spark, or a ½ K.W. close core transformer.

Suppose your home is in Austin, Texas. The nearest state line is Louisiana, a distance of 230 miles. Thus you could with perfect safety use our 200 mile sending outfit, which does not reach more than 200 miles.

It is also pointed out that if you live within five miles of a Government wireless station you cannot use more than ½ K.W. power, though the next state border might be 100 miles or more distant. Of course if you live close to another state, as for instance, in New York City, you are required to take out a license for any size transmitter.

The license has not been created to muzzle you; it is the other way around. Uncle Sam gives you a written order telling you that you can send messages to your heart's content, **and no one can tell you to stop sending**, as long as you do not create mischief.

The license is free. It costs not a penny. All that is required of you is that you are familiar with the law and that you can transmit messages at a fair degree of speed.

> **What the License Is**

The law does not require that you take an examination in person if you are located too far from the nearest radio inspector. All you have to do is to take an oath before a notary public that you are conversant with the law and that you can transmit a wireless message. If you wish to be licensed—and we urge all amateurs to do so, as it is a great honor to own a license—write your nearest Radio Inspector (see below), and he will forward the necessary papers to you to be signed.

Radio inspectors are located at the following points: (Address him at the Customs House):

Boston, Mass., New York, N. Y., Baltimore, Md., Savannah, Ga., New Orleans, La., San Francisco, Cal., Seattle, Wash., Cleveland, Ohio, and Chicago, Ill. Also the Commissioner of Navigation, Department of Commerce and Labor, Washington, D. C.

In an interview with the New York Times, W. D. Terrell, United States Radio Inspector for the port of New York, said in discussing the new law:

"The new law regulating wireless messages will work no hardship to the amateur operator. It is the intention first, to classify the various operators and place each operator in his proper class. They will then be permitted to work or play as much as they please, but under an intelligent, general supervision. Only those stations are affected which are near enough to the coastal stations to offer interference, or which work across the state lines which brings them under the supervision of the inter-State laws. I would like to make it very clear that the license costs the amateur nothing, and that the Government is willing to facilitate the wireless operators in every way possible to secure their license."

So much for the law. Everybody will now understand that the law is just and fair and that it gives the amateur a distinct standing in America, a standing which he does not enjoy in any other country. He knows what he can do and what he can't do, and no one can come to him and boss or abuse him, as Government or Commercial wireless operators were wont to do before the enactment of the law.

With sending outfits the reasoning is about the same as with the receiving outfits.

In order to select an outfit you must of course know where and how far you wish to send. Upon this, all depends.

| Sending Outfits |

As a rule—9 out of 10 of our customers have done it—two or more friends get the "Wireless bug" and order two or more complete transmitting sets. Of course, the outfits selected must necessarily be powerful enough to cover the intervening distance between the houses of the friends, and this only you know.

Therefore if you and your friend decide to converse by wireless and if the distance between your two houses is 10 miles you will probably buy our 15 mile sending outfit. Of course, a more powerful set may be used, although there is no particular advantage in doing so, except perhaps that the incoming signals of necessity will be louder with the more powerful sets. It goes without saying that almost ANY receiving outfit which we list can be used with ANY of the sending outfits. Bear in mind that the selections which we give with our sending outfits do not have to be used if not wanted. Thus, our "Interstate" outfit or even our "Nauen" receiving outfit can be used with our 3 mile sending outfit. For if you and your friend live two and a half miles apart and both of you have 3 mile Sending Outfits, you probably want to have a receiving outfit with which both of you can pick up messages 2,000 miles distant. In that case you would order two 3 mile Sending Outfits only, and two 1,500 mile receiving outfits, or else two "Nauen" receiving outfits. If either you or your friend feel that you cannot afford such a set, why then get the set that you can afford best and that suits you best. As you see there is no hard and fast rule about the relation of sending and receiving outfits. On the other hand we don't have to tell you that if you order two 200 mile sending outfits you require of necessity a good receiving outfit, else you couldn't hear the station 200 miles off. A little common sense will help everyone decide just what combination to order.

Like receiving sets, the transmitting sets are divided into two groups. The un-tuned (open circuit) and the tuned (closed circuit) ones.

The untuned ones have, 1st—a spark coil, 2nd—source of power, usually dry cells or a storage battery, 3rd—the spark gap. 4th—the key.

Sending Apparatus

Such outfits can be used only for very short distances and should never be used above three miles. When connections are made by following the blue prints, which we supply with all sets, the pressing of the key gives a strong spark in the spark gap. The spark gap (the open space between the zinc plugs) from the smallest to the largest sets, must never be more than one-eighth to three-sixteenths of an inch. A bigger gap does not work. Pressing the key long gives a dash, pressing it but for a fraction of a second gives a dot. Combinations of these represent the telegraphic characters; the code can be learned in a few weeks, practicing twice a day from one-half to one hour. (See Lesson No. 15, of The Wireless Course.)

In the tuned outfits, we have in addition to the above enumerated apparatus: 5th—The Leyden jars, or condenser; 6th—The Helix, or oscillation transformer. The Leyden jars change the red spark obtained from a spark coil, into an intense blue-white crashing spark. The Leyden jars also create a train of fast oscillations and go to make the outfit far more powerful although no more battery power is required. The Leyden jars also give better "carrying power," as the signals can be heard more distinctly and not "mushy" as if no Leyden jars were used. For each outfit the best jars or condenser has been selected and no changes should be made here.

The helix as well as the oscillation transformer, are, to the sending outfit, what the tuner and the loose coupler respectively, are to the receiving outfit. The helix or the oscillation transformer is the tuning coil pure and simple for the transmitting station. Like the tuning coil the helix and the oscillation transformers have sliders or else clips by means of which more or less wire convolutions can be put in the circuit of the aerial. Therefore more or less wire, and consequently more or less wave length is added to your aerial. Again there is not much of a mystery here. Anyone understands it. (See Wireless Course Lesson No. 14.)

In the larger sets where the battery power is insufficient as well as un-economical we have two methods open to fill the gap. One is the Gernsback electrolytic interrupter working on 110 volts Direct or Alternating current, which supplies the spark coil (transformer coil) with the power; the other method requires the use of a CLOSED core transformer operating without any kind of interrupter direct from the alternating current supply. This kind of transformer, however, does not work on the direct current, not even in connection with the electrolytic interrupter. The choice, for this reason, lies entirely with you.

The aerial switch is an absolute necessity where both a sending and receiving set is used in one station. If you are through receiving a message from your friend, you, of course, wish to answer him. You therefore, must switch the receiving set off from your aerial and switch the sending onto the aerial. The aerial switch does all this in one operation.

For sets using nothing higher than a 2½-inch spark coil an ordinary double pole, double throw switch may be used. For heavier sets using more power our Antenna switch No. AEK8100 must be used, as the smaller switch cannot carry the necessary power. We furnish complete diagrams for the connections.

In order to send messages it goes without saying that you must know how to "tap the key." The easiest way to learn, and the cheapest way at the same time, is to get a buzzer set as explained under "Reception of Messages." With this set, which represents a first class learner's outfit, you can send yourself dots and dashes to your heart's content until your wrist has limbered up sufficiently to do rapid sending. After a few weeks' practice it will be as easy to send a telegraphic message as to write on a typewriter.

Sending a Message

If your friend has a wireless and starts learning the code with you, it becomes very simple for both of you to soon become proficient in the art. Each will send to the other, the Morse or Continental alphabet, which is sent back and forth till the right speed is obtained. After this certain words are exchanged between the stations; later on short sentences are sent and so forth, till it becomes possible to converse freely by wireless.

There is but little adjusting to do when sending. As a rule amateurs converse with only one, seldom two, and rarely three stations. For this reason much adjusting is unnecessary. When using a small set comprising Spark Coil, Leyden jars and helix it becomes first necessary to adjust the Leyden jars. Either more or less jars (which adds more or less capacity to the circuit) are used till the spark sounds loudest in the spark gap and appears most powerful. A little experimenting will quickly tell when the right capacity is used. **It is important to understand the capacity should be adjusted only when the spark gap is connected to aerial and ground.** (See Lesson No. 14, Wireless Course.)

The next important adjustment is in the helix (or oscillation transformer if this is used in place of a helix). To change the clips around on the helix (or on the oscillation transformer) it is necessary that a small gap is first made in the aerial circuit. This is done best by driving two nails in a piece of very dry wood, and connecting the aerial wires to each nail as shown in sketch. The two wires A and B are brought close together now and when the key is pressed down a small spark will jump from A to B showing that you are charging the aerial and that energy is radiated from same. Now change the adjustment on the helix (or oscillation transformer) till the longest and fattest sparks jump between A and B. To do this A and B are separated until a point is reached where the spark cannot jump any further. You know now that you are radiating the maximum of energy and the point on the helix (or adjustment on the oscillation transformer) should be carefully marked so you will know at any time just where the maximum is. It goes without saying that you should also note how many Leyden jars (or how many condensers) you are using when making the test and you should write this information down, for if you were to use more or less Leyden jars (or condensers) you would have to change the adjustment on the helix (or oscillation transformer) as explained above. Now after the maximum "radiation" has been ascertained, the test block with the nails is discarded and the break in the aerial wire connected again. You know now that your station is radiating the maximum energy and adjustments of the sending set will not be required for some time to come. Indeed they may be left undisturbed indefinitely.

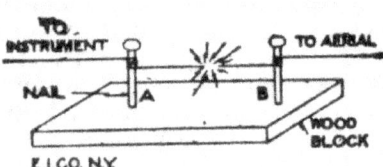

We believe that we have made everything as plain as possible and that by reading this treatise the elementary points of "Wireless" must become plain to even the layman. If, however, you desire additional information, our engineering staff will be only too glad to answer your questions promptly and explicitly. Now it's up to you to get busy and "start something"!!

Now suppose you have ordered your instruments from us and have erected your aerial. For the first few weeks you will be thrilled as you daily receive the far distant stations not only from all over the country but from far distant Germany, providing of course that you have the correct instruments for doing so. Sometimes also you will catch wireless telephone messages, as more and more Radio telephone sets come into use every day.

Wireless Clubs

Soon will come the time when you wish to chat with your friends by wireless. They will see your station and will be so impressed that they will want to have one of their own. Then why not start a local Radio Club in your town yourself, and become its president and founder? Indeed nothing is simpler. But first you must belong to a National Body and this you will find in the RADIO LEAGUE OF AMERICA, the biggest association in the country to which every amateur of note belongs.

The Radio League of America is a big scientific institution and ANYONE who has a wireless station can and should belong to it. It is a non-money-making institution; there are no fees and no dues to be paid. The League moreover furnishes every member with a free 'certificate printed in green and gold. Each member's station will, moreover be registered free of charge at Washington in the Government records, so that in case of war, every patriotic amateur can help his country in important scout work.

The League has the following distinguished members, of world renowned fame:

>Captain W. H. G. Bullard. U. S. N.
>Dr. Lee de Forest.
>Professor Reginald A. Fessenden.
>Nikola Tesla.

The affairs of the League are managed by Mr. H. Gernsback.

You should find an 8-page booklet telling you all about the League with this catalog. In case you did not get it or if you want more for distribution among your friends, write to

>RADIO LEAGUE OF AMERICA,
>233 Fulton Street, New York City.
>No charge is made for this service.

Fac-simile of button

FINIS

Application for Membership in the Radio League of America

I, THE UNDERSIGNED, a Radio Amateur, am the owner of a Wireless Station described in full on the back of this application. My station has been in use since.................................., and I herewith desire to apply for membership in the RADIO LEAGUE OF AMERICA. I have read all the rules of the LEAGUE, and I hereby give my word of honor to abide by all the rules, and I particularly pledge my station to the United States Government in the event of war, if such occasion should arise.

I understand that this blank with my signature will be sent to the United States Government officials at Washington, who will make a record of my station.

Witnesses to signature: Name...............................

....   City...............................

State...........................

....   Date..................191

In the event of national peril, will you volunteer your services as a radio operator in the interest of the U. S. Government?..............

This last question need not be answered unless you desire it.

Cut out, fill out blank and mail to:
>Manager Radio League of America, 233 Fulton St., New York.

Receiving Time by Wireless

(Reprinted from "The Electrical Experimenter")

The wisdom of furnishing vessels at sea with the correct time by wireless has been demonstrated time and again since the government began sending the signals over a few years ago. Since that time, too, many jewelers, railway officials and others on land who need the correct time have been taking the messages. The sending instruments used are extremely powerful, and any owner of a wireless receiving outfit within their range may get the time absolutely correct twice a day by properly tuning his receiving apparatus.

The stations that send out the reports, and the wave lengths used by each, are as follows:

Arlington (NAA)—2,500 meters.
Key West (NAR)—1,000 meters.
New Orleans (NAT)—1,000 meters.
North Head (NPE)—600 meters.

Eureka (KPM)—600 meters.
San Diego (NPL)—600 meters.
Tatoosh, Wash. (NPD)—600 meters.

The time signals have been easily and distinctly noted over two thousand miles from the sending station; and as the waves travel one hundred and eighty-six thousand miles a second, the difference between the sending and receiving time is practically nil.

All the stations send the time signals at noon and ten p. m., the first three according to seventy-fifth meridian time and the last four according to one hundred and twentieth meridian time. Our "Tuckerton" or our "Nauen" receiving sets are highly recommended for time signal work. The "Tuckerton" or the "Nauen" set is supplied complete already wired in cabinets. All that is necessary for the purchaser to do, is to erect the aerial, and connect the aerial and ground wires to the Binding Posts marked AERIAL and GROUND on the instrument. The wireless time signals are sent out from the United States Naval Wireless Telegraph Stations located at the Navy Yards throughout the United States; and the form of time signal is as follows:—For the five minutes preceding 12 o'clock noon a series of short dots, manifested as short buzzes in the Wireless Receivers, are sent out about one second apart; and at exactly 12 o'clock the series of signals end with a short dash, and this dash starts exactly at 12 noon. Hence when the start of this final code dash is heard in the receivers, it is exactly 12 o'clock. The time duration of this dash is about 1.5 seconds long, and dots are about .35 seconds long.

The transmitting clock that mechanically sends out the signals is corrected very accurately, shortly before noon, from the mean of three standard clocks, that are rated by star sights with a meridian transit instrument.

We recommend for this purpose, either our "Tuckerton" Receiving set or our type "Nauen" Receiving set.

The aerial for this purpose should be quite large and preferably have a height above the ground of not less than 75 to 100 feet. The aerial itself may be of the flat-top or slanting variety; and may be composed of 6 to 8 strands of our solid Antenium wire or stranded Antenium cable, each strand having a length of 80 to 100 feet, and for very long distances such as 1,000 miles or more, the aerial should be as high as possible, and probably 150 to 200 feet in length, or even more. The strands may be spaced 5 to 6 feet apart.

Arlington—Time Signals 2500 Meters

Note. The Arlington Station sends time signals on a wave length of 2500 meters commencing at 11:55 A. M. and 9:55 P. M. every day. Final signals at 12 Noon and 10 P. M. are for the meridian of 75 degrees west of Greenwich. Every tick of the standard clock of the Naval Observatory, Washington, is transmitted as a dot, omitting the 29th second of each minute, the last five seconds of each of the first four minutes and finally the last ten seconds of the last minute. The 12 Noon and 10 P. M. signal is dash.

"Inter-Ocean" Wireless Receiving Outfit

The increasing popularity of loose coupled wireless receiving apparatus has caused us, as leaders in the manufacture of amateur wireless equipment, to design an outfit which, at its price, is positively unequalled at the present time.

It consists of our latest model loose coupler No. DBE 12002 with single slide, bare wire wound primary; with a secondary of silk covered wire. A switch handle is provided on the secondary to permit of the closest possible tuning adjustment, so important in long distance work. Signals can be perfectly received from stations using long wave lengths, for our celebrated No. BEK8487 Loading Inductance is provided giving a receiving capacity **up to 6,000 meters wave length.**

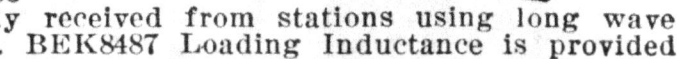

No. HHE1550

The detector is our No. ABE9700 Galena Detector which for sensitivity, convenience and permanence of adjustment is positively unsurpassed at the price. Our No. FK10010 Jr. fixed condenser is also part of the outfit and is a big aid in making long distance signals audible. The entire outfit is wired at the factory and is mounted on a finely finished oak base with nickel plated binding posts for aerial ground and phones. Every binding post is identified by a neatly etched brass name plate. Read the detailed descriptions of the individual apparatus (described elsewhere in this catalog), comprising this outfit, and you will better appreciate its great value.

No receivers are furnished with the "Inter-Ocean" outfit but excellent results can be obtained with any of our head sets. We especially recommend our No. GX6666 Government Receivers with which we have very clearly heard stations 500 miles away using an aerial only 100 feet long by 50 feet high. The "Inter-Ocean" outfit is something you will never regret buying and one we are proud to mark as a product of the E. I. Co. Size 10x12x6½ in. high.

No. HHE1550 "Inter-Ocean" Wireless Receiving Outfit, complete as described, but no phones.......................... **$8.85**
Shipping weight 10 lbs.

No. AAHE1551 "Inter-Ocean" Outfit, with No. CX8090 "Junior" Wireless Phones **$11.85**
Shipping weight 12 lbs.

No. ABHE1552 "Inter-Ocean" Outfit, with No. DX8070 "Amateur" Wireless Phones........................... **$12.85**
Shipping weight 12 lbs.

No. ADHE1553 "Inter-Ocean" Outfit, with No. FX1305 "Trans-Atlantic" Wireless Phones **$14.85**
Shipping weight 12 lbs.

No. AEHE1554 "Inter-Ocean" Outfit, with No. GX6666 "Government" Wireless Phones **$15.85**
Shipping weight 12 lbs.

Permit us to let us send you FREE, with our compliments when ordering any of our Receiving apparata, lesson No. 1 **"The Principles of Electricity"** or lesson No. 9 **"The Receiving Apparata"** or lesson No. 14 **"The Operation of the Instruments"** or lesson No. 12 **"The Hook-Ups and Connections"** of our famous "WIRELESS COURSE."

The Principle and Construction of Receiving apparata are described in these lessons.

Just attach your free coupons to your order. For further information consult colored section in this catalog.

The "Interstate" Wireless Receiving Outfit

No. DBE1500

This is the outfit you have been looking for. For the money it certainly eclipses everything that has been offered heretofore in this line. IT IS NOT A TOY by any means, and while we do not claim it to be a commercial set, we do claim that the amateur who owns one is enabled to do excellent work in all respects. On our house aerial, in connection with this set, we have distinctly heard the Boston Navy Yard, 300 miles distant, and were enabled, by means of the tuning coil, to "cut out" the very strong interference from the powerful station on the Woolworth Building, a quarter of a mile away from us. We therefore do not think it necessary to add further praise to this outfit.

The very best instruments are used in the "Interstate." We do not believe in throwing things together, just to sell the outfits. We don't save pennies, in order to make you spend dollars. If you don't like the outfit, return it. That's our policy. Therefore we must build it right in the first place.

This set comprises:
Our No. BX9950 Standard Tuning Coil, bare wire wound, with two patent sliders, hard rubber composition coil head and posts; our new improved **GALENA DETECTOR** of wonderful sensitiveness, **ONE OF THE MOST SENSITIVE IN EXISTENCE;** our No. EK1024 75 ohm Watch Case Receiver and 3-foot cord; a good condenser, mounted in base, balanced to the right capacity for this set, and a ½ inch thick **SOLID HARD RUBBER COMPOSITION BASE** on which all the instruments are mounted. A handy nickel hook is also provided, on which to hang the receiver when not in use.

ALL CONNECTIONS ARE MADE. When you get the outfit it is ready for use. Just attach the aerial to post 1, and ground to post 2, and the "Interstate" is ready for use. Full and very explicit directions go with the outfit.

In order to receive messages up to 300 miles or over, a 4-wire aerial about 50 feet above ground, 50-75 feet long is required. For this you need 200 or 300 feet of our No. DE9219 Antenium wire and 8 of our No. AF10007 insulators.

By using the No. AGE8071 Receiver the distance is increased from 20-25 per cent. With the No. BEK1307 Receiver still better results are obtained.

Size over all of this set is 9x6x6 inches.

No. DBE1500 "Interstate" Wireless Outfit, as described.......... **$4.25**

No. EEK1501 "Interstate" Wireless Outfit, with No. AGE8071 Receiver **$5.50**

Shipping weight either DBE1500 or EEK1501, 4 lbs.

NOTE.—By using our No. BEK8487 loading coil in series with the tuning coil the wave length of this outfit is increased greatly, and time signals from the powerful Government stations can be plainly heard.

Gentlemen:—   Minneapolis, Minn.

We are using some of your apparatus and THEY WORK GOOD. We have received some of the radio stations on the coast and other large stations such as Arlington, Sayville and others.

Yours truly,   C. BOOTH.

The "Trans-Pacific" Wireless Receiving Outfit

No. AKX1555

Just as its name implies so is the outfit. It is the outfit for long distance, truly great distances.

Simplicity, efficiency, quality and low price. These four words were constantly before us when this outfit was being developed. Your first test of the "Trans-Pacific" will show you conclusively how carefully we have adhered to our ideals.

The No. GEK14000 Professional Loose Coupler, with a wave length of 4,000 meters, permits of its use for receiving all the commercial stations. The coupler itself is the simplest and most beautiful instrument of its kind we have ever developed, and that is saying much. The "Universal" detector has been on the market many years and its rotary sliding cup (patent applied for) and double micrometer and spring adjustment are really the acme of detector simplicity and efficiency. The blocking condenser is not the ordinary fixed kind, but is our No. ABE10000 Fixed-Variable type which gives 3 capacities to work with, enabling a choice of the one which will bring in the signals loudest. The entire outfit is superbly finished and mounted on a finely polished base. All wiring if done at the factory and the outfit comes ready for connections to an aerial and ground. There are four nickel-plated binding posts, each appropriately marked by a neat etched brass metal name plate. No receivers are furnished with the "Trans-Pacific" outfit but we particularly recommend our No. FX1305 Trans-Atlantic Phones, or No. GX6666 Government Phones. This outfit has received messages over 700 miles with a 100-foot aerial 50 feet high with the use of our "Government" phones and it will repeat the work for you at any time.

Made by the pioneers in the manufacture of amateur wireless equipment, this outfit looks and works as though it had every bit of knowledge we had acquired in 12 years put into its remarkable value. Size 8x20x7½ in. high.

No. AKX1555 "Trans-Pacific" Wireless Receiving Outfit, complete as described, no Phones..................... **$10.00**
Shipping weight 15 lbs.

No. AFX1556 "Trans-Pacific" Outfit, with our No. FX1305 "Trans-Atlantic" Phones **$16.00**
Shipping weight 18 lbs.

No. AGX1557 "Trans-Pacific" Outfit, with our No. GX6666 "Government" Phones **$17.00**
Shipping weight 18 lbs.

Gentlemen:— Kenosha, Wis.

The instruments arrived a few days ago and after giving them a pretty thorough test, I think I can justly say, THAT THEY ARE SATISFACTORY IN EVERY RESPECT.

Let me say to whom it may concern, that this statement is absolutely unsolicited on the part of the ELECTRO IMPORTING CO., being entirely voluntary on my part. Yours respectfully, SIDNEY DERBYSHIRE.

The "Electro" Intercity Sending Outfit

DESIGNED ESPECIALLY FOR BOY SCOUT SERVICE

NO. GEK1900

While we make quite a few styles of complete receiving outfits which are all fully connected, the "Intercity" outfit is the only sending outfit we make that is all completely connected ready for service.

Every bit of wiring is done at our factory and as only standard stock articles are used, it is a certainty that the outfit must give a maximum of results even considering its low price.

The outfit comprises one of our well known No. DEK1088 one inch Bull-dog Spark Coils, one No. EK9220 Spark Gaps, one No. ABE1117 Telegraph Key and one special No. GE2345 Sending Condenser. All the apparatus is carefully mounted on a finely finished heavy oak base in the bottom of which the condenser is sealed.

In operation the outfit is simplicity itself. Just connect aerial and ground wires and your battery and you are ready for work. The outfit has a range of 5 miles when used on an aerial 50 ft. long and 40 ft. high. This range can be increased by the use of a larger aerial. As it has no helix or oscillation transformer it does not send out tuned waves, but any one of our helices or oscillation transformers will work perfectly on this outfit.

We do not hesitate to recommend this outfit to the most particular customer for we know what excellent labor and materials we put in it and know how well satisfied every customer has been who has bought one, and many thousands have been sold in the last 5 years.

No. GEK1900 "Intercity" Wireless Sending Outfit, complete as described **$7.50**
Shipping weight 12 lbs.

No. AKGE1901 "Intercity" Wireless Sending Outfit, complete as described, but with the addition of aerial outfit No. BX333, consisting of 125 ft. No. DE9219 Antenium Wire, one No. EK3339 Connector and five CK1001 "Electro" Dry Cells. This makes an absolutely complete sending outfit. Price complete **$10.75**
Shipping weight 15 lbs.

No. AEIK1902 "Intercity" Sending Outfit (5 mile range) also No. CGE1500 "Interstate" Receiving Outfit (range 300 miles) also our No. BX333 Aerial Outfit, and our No. DK1313 Switch, and 5 No. CK1001 Dry Batteries. A wonderful outfit for all Boy Scout Organizations for whom it was especially designed. Price complete, as described **$15.90**
Shipping weight 25 lbs.

All these outfits whether No. GEK1900 or No. AEIK1902 are supplied with code chart of Morse, Continental and Navy codes, also complete directions for erection, care and use of the outfits purchased.

The "Electro" "Arlington (NAA) Baby Timer"

De Luxe Receiving Cabinet

Wave Lengths: Min. 200 Meters; Max. 1,200 Meters

No. HEK4433

The "Arlington Baby Timer" is without question the most compact, the smallest, as well as the most wonderful little time receiving cabinet on the market to-day.

With a fairly large aerial, it will receive stations such as Arlington (NAA) over a distance of 1,000 miles with ease. For a little outfit it has no peer in selectivity.

Very fine selectivity is had both with condenser and tuning inductances. The tuning is sharp and accurate and you will be amazed at the clearness of the received signals.

In this outfit a standard Auto Transformer type of tuning inductance is used. This type makes for great selectivity as practically no energy is lost in the transformation. This makes it in a sense more efficient than most loose couplers on the market to-day.

The outfit acts similar to an interference preventer, because the variable condenser is especially connected for efficient selective tuning.

The "Arlington Baby Timer" is recommended for use with an oscillating vacuum detector for undamped waves if external connections are made as per diagram furnished with the cabinet.

It can of course be used with any detector and any good set of phones.

FULL BAKELITE FRONT

As in all our **De Luxe** sets, the entire front of the outfit is made of BAKELITE. This material is now used in all high grade commercial outfits. It is much more expensive than hard rubber. It has these advantages: It has a beautiful black finish which it never loses. Unlike hard rubber, it never warps. It is the best electrical insulator known to science to-day, barring none. It does not collect, condense or absorb moisture—a very important point to consider in a wireless receiving cabinet. **All our binding posts, switch-points, all wiring, etc., are directly mounted on the Bakelite Panel.** This is the very best as well as the most expensive method known to science to-day. But when you want a Radio Cabinet, you want the best. Bakelite is THE best. We can't afford to use screw contacts so we solder every connection. It's expensive but it makes a better outfit.

The variable rotary Condenser used is our famous 17 plate No. BEK9240 type. None better can be made. The cabinet is of mahogany, with a hand rubbed piano finish. All workmanship is of the highest grade throughout. All metal parts heavily nickel-plated and hand buffed. Six generous nickel binding posts and four etched name plates are furnished.

Explicit diagram showing hook-up as well as various clever connections is furnished with the outfit.

By placing a loading coil in the aerial circuit, longer waves can of course be received with this outfit.

For a low priced radio cabinet outfit, this one is unmatched. Our illustration is but a poor attempt to convey an adequate picture to you. You must see and try this outfit—words cannot describe it.

Size over all 8¾x6x2 inches. Shipping weight 4 lbs.

No. HEK4433 "Electro" "Arlington (NAA) Baby Timer" as described **$8.50**

The "Electro" "Key-West (NAR)" Radio Outfit
DeLuxe Receiving Cabinet
Wave Lengths: Min. 200 meters; Max. 2,000 meters

NO. ADX4444

Here is a high grade long wave, long distance outfit. Will receive messages from large stations such as Key-West (NAR) Florida, over a radius of 1,250 miles on a medium aerial. Longer distances with a larger aerial. Arlington (NAA) time signals can be copied perfectly with either a good crystal or vacuum detector.

For a high grade medium priced outfit, this cabinet cannot be surpassed. It stands in a class by itself.

Wonderful selectivity is assisted by the 43 plate rotary variable condenser, which is shunted across the secondary. This condenser, by the way, has a most decided influence upon the whole system and the tuning can be controlled very accurately by its means. For very fine tuning, switch selectors No. 2 and No. 3 are used.

In this outfit a closely coupled tuning inductance of highest selectivity giving peculiar qualities is employed. The inductances are wound on special tubes and their balance is so perfect that great selectivity is guaranteed. On account of the close coupled transformer used, practically no energy is lost in the transformation, thus making the outfit more efficient than most loose couplers on the market.

As the tuning is very sharp most interference can readily be eliminated. The outfit therefore in many respects is a fine interference preventer.

Our "Key-West NAR" outfit can be used with any kind of crystal or vacuum detector. We also furnish a diagram showing how undamped waves can be received with this cabinet, using an oscillating vacuum detector.

FULL BAKELITE FRONT

As in all our De Luxe sets the entire front of the outfit is made of BAKELITE. This material is now used in all high grade commercial outfits. It is far more expensive than hard rubber. It has these advantages: A beautiful black finish which it never loses. Unlike hard rubber, it never warps. It is the best electrical insulator, known to science today, barring none. It does not collect, condense or absorb moisture—a very important point to consider in a wireless receiving cabinet. **All our binding posts, switch-points, and all wiring, etc., are directly mounted on the Bakelite Panel.** This is the very best as well as the most expensive method known to science to-day. But when you want a Radio cabinet, you want the best. Bakelite is THE best. Furthermore, every connection is soldered, not simply a screw contact.

The variable rotary Condenser used in our famous 43 plate **No.** DX9241 type. None better can be made. The cabinet is of mahogany.

with a hand rubbed piano finish. All workmanship is of the highest grade throughout. All metal parts are heavily nickel plated and hand buffed. Six generous nickel binding posts and seven etched name plates are furnished.

Explicit diagram showing hook-up as well as various clever connections is furnished with the outfit. By placing a loading coil in the aerial circuit, longer waves can of course be received with this outfit.

This very handsome Radio outfit cannot be matched anywhere at our price. It is not only very compact, but is readily portable as well. It looks more like a $30.00 outfit; with its expensive Bakelite front, once it is in your station. It is a little beauty all the way through and our illustration does not do it justice whatsoever. You must see the "Key West" to fully appreciate it.

Size over all 8x8x4⅜ inches. Shipping weight 6 lbs.

No. ADX4444 "Electro" "Key-West NAR" Radio Outfit as described **$14.00**

The "Electro" "Sayville (WSL)" Radio Outfit
De Luxe Receiving Cabinet
Wave Lengths: Min. 450 Meters; Max. 2,500 Meters

Our "Sayville (WSL)" outfit was originally planned for jewelers' use. As can be readily understood a jeweler's outfit must be of commercial type, must be highly selective and practically free from interference.

All this and much more is accomplished with this efficient set.

Time signals from such stations as Arlington (NAA) are received with the greatest ease over a radius of 1,500 miles on an aerial having 310 meters. This is accomplished with any good crystal or vacuum detector. On larger aerials the receiving distance of course is increased.

No. AHX4455

Very sharp tuning is had particularly with the center switch selector of the primary transformer winding. The two, 43 plate, rotary variable condensers are particularly useful for tuning waves around 2,500 meters in length. THE CONDENSER AT THE RIGHT HAS A SHORT CIRCUITING ARRANGEMENT FOR INCREASING THE WAVE LENGTH. WHEN OPEN THIS CONDENSER REDUCES THE WAVE LENGTH AND AMATEUR STATIONS CAN THEN BE RECEIVED WITH EASE.

In this outfit we use very efficient, closely coupled tuning inductances AND A SPECIAL VARIOMETER COIL placed at right angles to the tuning inductance. This construction is original with us and the entire arrangement of closely coupled inductances is more efficient than most loose couplers, as practically no energy is lost in transformation.

The "Electro" "Sayville (WSL)" Radio Outfit
(Continued)

This outfit is highly selective especially on long wave lengths. It is an ideal jeweler's set and the better, up-to-date amateur will be proud of this fine set. Four binding posts for the phones are used in order that two sets may be attached to the outfit.

By means of the three switch selectors nearly all interference can be eliminated, which is especially important when receiving time signals.

Our "Sayville (WSL)" outfit can be used with any kind of detector either crystal or vacuum type. A diagram is furnished showing how undamped waves can be received with this outfit by using an oscillating vacuum detector.

FULL BAKELITE FRONT

As in all our **De Luxe** sets, the entire front of the outfit is made of BAKELITE. This material is now used in all high grade commercial outfits. It is far more expensive than hard rubber. It has these advantages: It has a beautiful black finish which it never loses. Unlike hard rubber, it never warps. It is the best electrical insulator known to science to-day, barring none. It does not collect, condense or absorb moisture—a very important point to consider in a wireless receiving cabinet. **All our binding posts, switch points, all wiring, etc., are directly mounted on the Bakelite Panel.** This is the very best as well as the most expensive method known to science to-day. But when you want a Radio Cabinet, you want the best. Bakelite is THE best.

To insure the maximum of efficiency we solder every joint and contact. This means longer range and a guarantee of long life for your outfit. We don't use screw connections anywhere.

The two variable rotary condensers used are our famous 43 plate No. DX9241 type. None better can be made. The cabinet is of mahogany, with a hand rubbed piano finish. All workmanship is of the highest grade throughout. All metal parts are heavily nickel-plated and hand buffed. Eight generous nickel binding posts and seven etched name plates are furnished.

Explicit diagram showing hook-up as well as various clever connections is furnished with the outfit.

By placing a loading coil in the aerial circuit, longer waves can of course be received with this outfit.

This highly efficient set stands unmatched in this country to-day. It is of generous proportions and represents a good deal more than what we ask you to pay for it. It looks business and will give your station a commercial looking appearance. Our illustration does the handsome set but scant justice. You can't adequately show mahogany, nickel and Bakelite in a black and white illustration.

Size over all 12x10x5½ inches. Shipping weight 9 lbs.

No. AHX4455 "Electro" "Sayville (WSL)" Radio Outfit as described **$18.00**

When ordering one of our Tuners, Loose Couplers, Loading Coils, Receiving Cabinets, etc.; permit us to present you **free** with our compliments the following lessons of our famous "WIRELESS COURSE": lesson No. 4 **"The Principles of Wireless Telegraphy"** or lessons Nos. 8 and 9 **"Receiving Apparata"** or lesson No. 12 **"The Hook-Ups and Connections"** or lesson No. 14 **"Operation of Instruments."**

You will learn how to tune your station properly and how to get the most out of your instruments. Just attach your free coupons to your order. For further information consult colored section of this catalog.

The "Electro" "Tuckerton (WGG)" Radio Outfit

DeLuxe Receiving Cabinet
Wave Lengths: Min. 780 Meters; Max. 3,20. Meters

The need of a genuine amateur suited set has been long felt by the various manufacturers, and after a considerable amount of research and experimental work, on our part, our engineers have finally devised an outfit which will cover ALL the faults inherent in outfits manufactured heretofore. We are the only company which has succeeded in perfecting such an outfit and it will do everything that we claim for it.

This outfit while eminently suited for the better amateur class, is highly recommended for jewelers' use, for the reason that the set is exceedingly compact and fool proof to a very high degree.

Time signals from Arlington (NAA) as well as other stations sending out the time twice daily, are received with astonishing ease with this outfit, signals are received over a radius of about 2,500 miles on a medium size aerial, in connection with a "Radiocite" or a vacuum detector, or any other sensitive detector and a good set of phones. With very large and highly elevated aerials the receiving distance is of course increased.

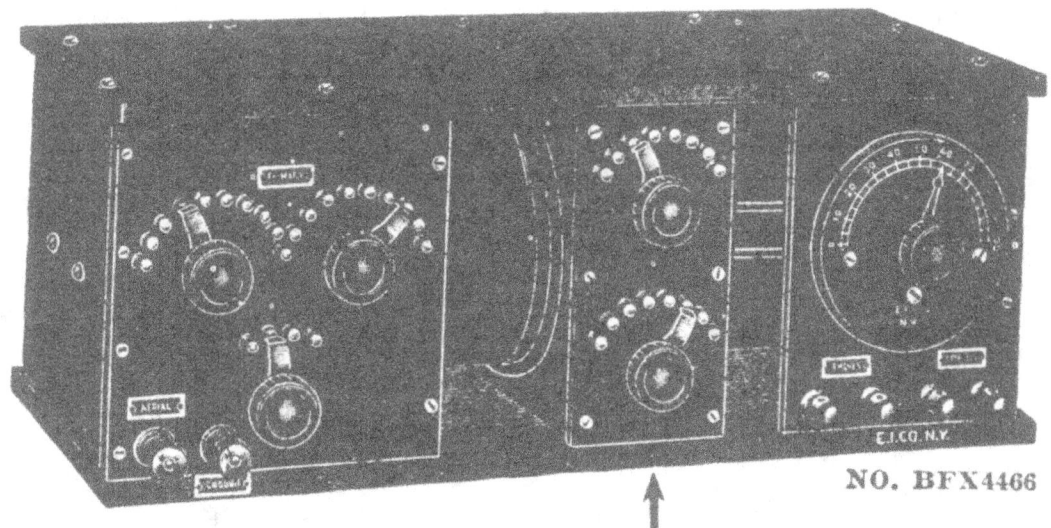

NO. BFX4466

SLIDES BACK AND FORWARD

This outfit is of the loose coupler type, especially balanced for fine and efficient tuning. The loose coupler is of the very latest approved type. The air space between primary and secondary is reduced to a minimum. Both coils are wound with silk covered copper wire—not enamel, the latter having proved inefficient for all fine tuning.

The primary is connected to two switch selectors, (shown at left in our illustration). The selector switch at left connects every tenth turn; it is used for coarse tuning. The selector switch at the right connects every turn of the first 24; it is used for very fine tuning.

A similar arrangement is used on the sliding secondary. Here two sets of switch selectors MOUNTED ON A BAKELITE PANEL are used; the top switch is for coarse tuning, the bottom one for fine tuning. When receiving, the switches of both primary and secondary are manipulated till the signals come in with maximum strength.

An additional switch is included with the outfit which short circuits the detector when the amateur wishes to transmit, so that the heavy currents produced by the transmitter do not affect the detector crystal. This switch is located below the two primary switches. The switch is a very essential part and has been overlooked in the past by ALL companies manufacturing wireless cabinets.

The Electro "Tuckerton (WGG)" Radio Outfit
(Continued)

For selective tuning the secondary should be loosely coupled with the primary by pulling it out from the primary and carefully tuning the secondary or left-hand condenser. It will be found that exceptionally fine tuning is obtained with this arrangement when once experience is obtained in the matter of handling the various parts of the outfit.

The secondary slides back and forward on two heavy nickel plated brass rods. It slides with wonderful ease and is pushed in or out of the primary by merely grasping one of the secondary switch knobs.

The outfit is so designed, and the circuit so arranged that it is possible to use it as a receptor for undamped waves. It is necessary to connect an additional loading coil in the aerial circuit to receive long waves, as stations which employ undamped wave generators use wave lengths exceeding 4,000 meters. Two of our No. DX8486 Tuning Coils can be connected in series and will work satisfactorily, in conjunction with the "Tuckerton" receiver for receiving wave lengths up to 6,000 meters, with a moderate size aerial. A suitable antenna for this kind of work should consist of one or two wires 300 to 600 feet long and about 50 feet high. Under normal conditions, with the above aerial and a properly tuned oscillating Audion, there should be no trouble in receiving European stations such as Nauen and Eilwiese with this wonderful outfit.

Our famous No. 9240 Variable Condenser is used. None better can be made. The cabinet is of **mahogany with a hand rubbed piano finish.** All workmanship is of the highest grade throughout. All metal parts are heavily nickel-plated and hand buffed. Six generous nickel binding posts and five etched name plates are furnished.

FULL BAKELITE FRONTS

As in all our De Luxe sets the entire front of the outfit is made of BAKELITE. This material is now used in all high grade commercial outfits. The switching panel of the movable secondary is of Bakelite too. It is far more expensive than hard rubber. It has these advantages: It has a beautiful black finish which it never loses. Unlike hard rubber it never warps. It is the best electrical insulator, known to science to-day, barring none. It does not collect, condense, or absorb moisture—a very important point to consider in a wireless receiving cabinet. **All our binding posts, switch-points, all wiring, etc., are directly mounted on the Bakelite Panels.** This is the very best as well as the most expensive method known to science to-day. But when you want a Radio cabinet, you want the best. Bakelite is THE best. Every connection is soldered. We don't use screw contacts anywhere.

Our "Tuckerton" (WGG) Outfit can be used with any kind of crystal or vacuum detector. We also furnish a diagram showing how undamped waves can be received with this cabinet, using an oscillating vacuum detector.

This very fine De Luxe cabinet is without a shade of doubt one of the greatest bargains in the country to-day. It has been designed for professional work and will stand a good deal of abuse. Its like will not be found anywhere. It is not a small set nor too large to be clumsy; it is just right, works right and keeps on working right, long after you have forgotten the modest price we ask for the set.

We are sorry that you can't see this outfit before buying it for our illustration does it a great injustice. How can we show the piano finished cabinet, the Bakelite fronts, the sparkling posts and the "works" in a lifeless picture? When you hold the cabinet in your hands and after you have tried it you will appreciate just how much we underestimated it. We can build outfits better than we can talk about them. Try us and see. Look at our testimonials, they talk better than we could.

Size over all is: 21½x7x6¾ inches. Shipping weight 15 lbs.

No. BFX4466 "Electro" "Tuckerton (WGG)" Radio Outfit as described **$26.00**

Dear Sirs:—   Hamilton, Ont., Can.
    I bought one of your D.S. tuning coils and two of your detectors and must say they work fine.   F. HOLIDAY, JR.

The "Electro" Commercial Wireless Receiving Outfit

Our long experience in quantity manufacture has enabled us to design a standard commercial receiving outfit which, for price, finish, durability, flexibility, tuning qualities and long range, is absolutely unsurpassed. It will tune to any wave length from 200 meters to 10,000 meters and the selectivity is so perfect that interference is absolutely impossible. Guaranteed under normal conditions and when used with an aerial 150 feet long and 100 feet high to receive 2,000 miles.

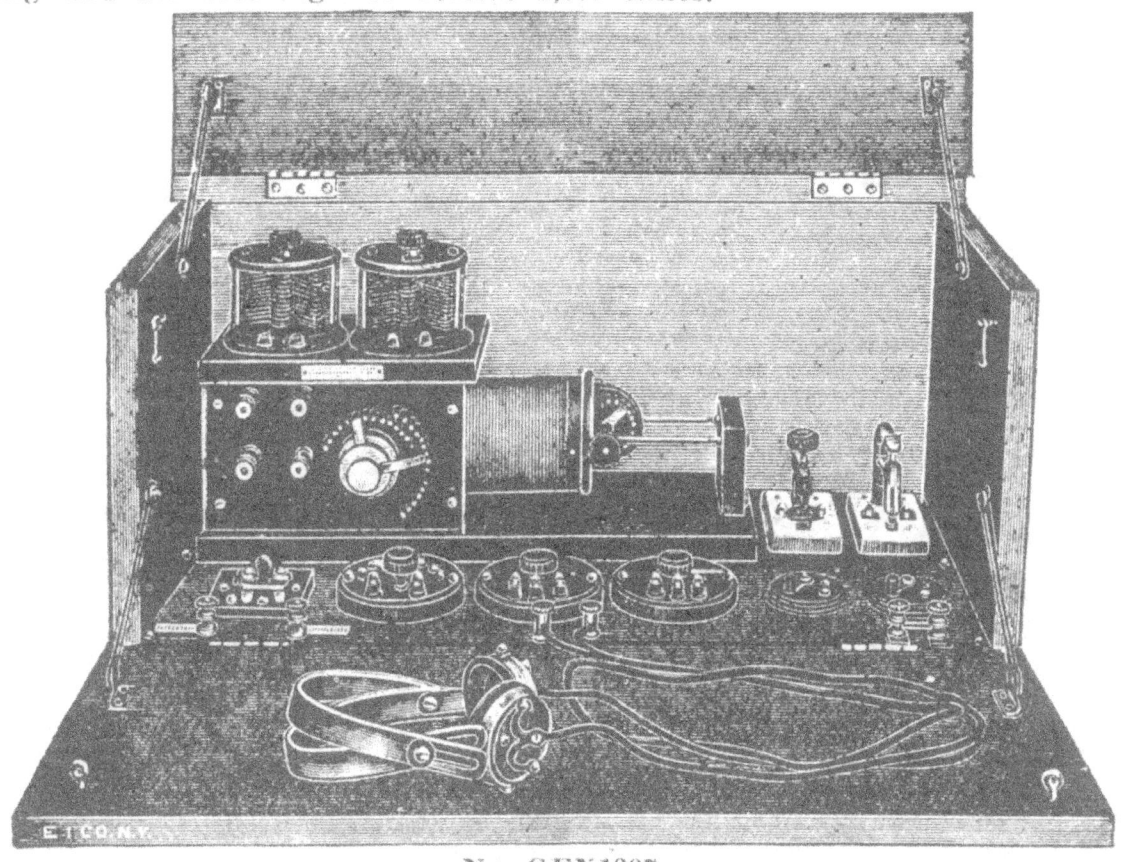

No. GEX1387

Case is of solid mahogany hand rubbed and finished throughout and closing entirely to make it dust proof. Convenient handles also make it portable. Coupler is of navy type of mahogany with switches on the primary, one of which controls the winding in groups of ten turns and the other in individual turns. The secondary has a switch of odd but efficient design. There is a 43 plate rotary condenser across the primary which can be disconnected at will by a switch. There is also a 43 plate rotary variable condenser across the secondary. For extra long wave lengths a loading coil with 7 points is provided. Two detectors, one a crystal type and the other the celebrated Radioson, ensure always one detector in good order. For the most perfect results with both detectors a rotary potentiometer and proper switches are provided as is a suitable blocking condenser. Needless to say all wiring is concealed and all switches are mounted on hard rubber and wherever possible of the rotary type. The binding posts are large, and well finished and all marked by nickel name plates. All metal is nickel plated and polished. One pair of 4,000 ohm "Government" receivers are furnished and of them nothing more need be said. Altogether it forms an outfit designed for commercial work, especially on ships. The outfit is absolutely guaranteed in every respect and of all those supplied by us not a single owner has made one complaint.

Cabinet size is 48x16x20 but the size can be slightly altered to suit special needs. Net weight is 50 lbs.

No. GEX1387 "Electro" Commercial Receiving Outfit as described **$75.00**
Shipping weight 80 lbs.

The "Electro" "Nauen (POZ)" Radio Outfit

De Luxe Receiving Cabinet

Wave Lengths: Min. 150 Meters; Max. 3,500 Meters

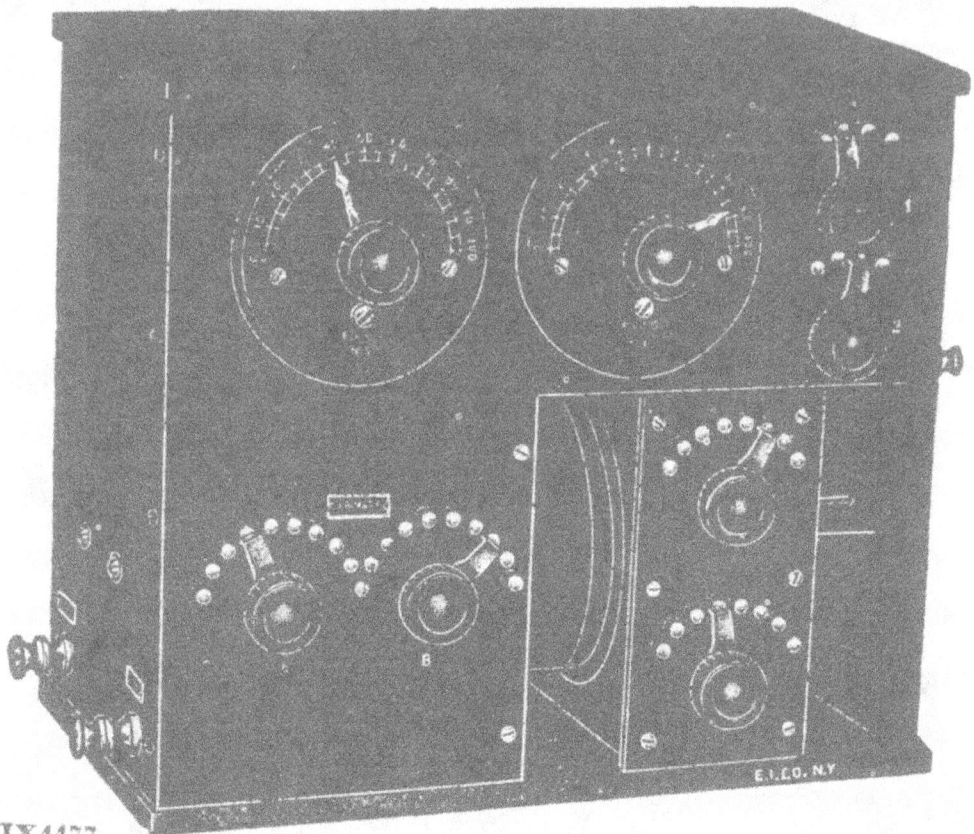

No. CIX4477

One of the finest designed radio outfits which was ever brought out by any radio concern is our "Nauen (POZ)" Radio Outfit which constitutes the most modern receiver.

This outfit is so designed that the circuit is capable of receiving both undamped and damped wave stations. It has a minimum wave length of 150 meters and a maximum of 3,500 meters which covers practically the entire range of modern radio practice.

Every "Nauen" outfit which we furnish is thoroughly tested in every respect; both for efficiency and operation. The outfit has a range of 3,500 to 4,000 miles with the use of a vacuum detector and a moderate size antenna. With a crystal rectifying detector, it is possible to use this set to receive stations 2,000 miles away, under almost all conditions, thus enabling the amateur to receive messages at all times which is impossible with most of the apparatus put out by other companies.

This is the only outfit which will stand severe usage, as the construction of the various parts of it is so rigid and durable that it may be handled in any manner. It is the only outfit really suited to club use where it is continually handled by the various members.

The cost of this outfit is so small as compared with usual high grade receiving sets, that it makes it applicable for ordinary radio club use, and we are certain that the good qualities inherent in this outfit will revolutionize the amateur field.

The outfit contains two sets of primary switches for controlling long wave reception for use in receiving continuous wave stations. In receiving long waves, switch blade No. 1 is turned towards the right, and the secondary is placed well within the primary. The secondary switches are placed on the extreme right hand switch point, thus using the complete winding of the secondary. The pointer of the left hand condenser should be set at 100 of the scale, thus giving maximum capacity to the outfit.

The "Electro" "Nauen" (POZ)" Radio Outfit
(Continued

In receiving short wave lengths, such as 600 meters and, below, switch No. 1 is re-set to contact on the left in order to connect the right hand condenser in series with the ground, thus decreasing the natural wave length of the outfit. The switches of both the secondary and primary are manipulated until the incoming signals are heard at the maximum intensity.

For selective tuning the secondary should be loosely coupled with the primary by pulling it out from the primary and carefully tuning the secondary or left-hand condenser. It will be found that exceptionally fine tuning is obtained with this arrangement when once experience is obtained in the matter of handling the various parts of the outfit.

This outfit is of the latest approved loose-coupled type. The secondary and the primary come as close together as is possible without touching each other. The secondary slides back and forward on two nickel plated thick brass slider rods. It slides with wonderful ease and is pushed in or out of the primary by merely grasping one of the secondary switch knobs. These switch selectors are mounted on a BAKELITE panel which slides back and forward with the secondary.

The outfit is so designed, and the circuit so arranged that it is possible to use it as a receptor for undamped waves. This is accomplished by an additional coil placed within the cabinet and forming part of the outfit. By merely throwing in the coil in the circuit by connecting in switch No. 2 to the right, it is possible to receive continuous wave stations. It is necessary to connect an additional loading coil in the aerial circuit to receive long waves, as stations which employ undamped wave generators use wave lengths exceeding 4,000 meters. Two of our No. DX8486 Tuning Coils can be connected in series and will work satisfactorily, in conjunction with the "Nauen" receiver for receiving wave lengths up to 6,000 meters, with a moderate size aerial. A suitable antenna for this kind of work should consist of a single wire 300 to 600 feet long and about 50 feet high. Under normal conditions, with the above aerial and a properly tuned oscillating Audion, there should be no trouble in receiving European stations such as Nauen and Eilwiese.

Two variable rotary condensers are used, the No. 9240 and our famous 43 plate No. DX9241 type. None better can be made. The cabinet is of mahogany with a hand rubbed piano finish. All workmanship is of the highest grade throughout. All metal parts are heavily nickel plated and hand buffed. Six generous nickel binding posts and four etched name plates, are furnished.

A diagram of connections is supplied with the outfit giving full details of connections and various hook-ups for use with the outfit.

Our "Nauen (POZ)" outfit can be used with any kind of crystal or vacuum detector. We also furnish a diagram showing how undamped waves can be received with this cabinet, using an oscillating vacuum detector.

FULL BAKELITE FRONTS

As in all our De Luxe sets the entire front of the outfit is made of BAKELITE. This material is now used in all high grade commercial outfits. **All our binding posts, switch-points, all wiring, etc., are directly mounted on the Bakelite Panel.** This is the very best as well as the most expensive method known to science to-day. But when you want a Radio cabinet, you want the best. Bakelite is THE best. Every connection is soldered. We don't use screw contacts anywhere.

This extraordinarily fine outfit, constituting the finest of our De Luxe sets, is head and shoulders above any Radio receiving outfit, selling at twice or thrice the price we ask for it. It is of generous design, but not clumsy by any means. It will make a station have that commercial look wherever used with a good pair of phones and a good detector.

We wish you could come to our factory and look at this outfit for the illustration conveys but little conception of the fine appearance of this our best outfit. You must see the "Nauen" to appreciate it.

Size over all 17½x11½x7 inches. Shipping weight 15 lbs.

No. CIX4477 "Electro" Nauen (POZ) Radio Outfit as described **$39.00**

"Electro" 300-Mile Receiving Outfit

A really excellent outfit that is dependable, tunable within a **very** wide range and remarkably flexible for so reasonably priced an **outfit.** The range of 300 to 500 miles is obtained with a 4 wire 'L" aerial 100 ft. long and 30 ft. high.

Outfit consists of:

1 No. DBE12002 Loose Coupler.

1 No. ABE9700 Detector.

1 pr. No. CX8090, 2,000 ohm Phones.

1 No. BEK8487 Loading Coil.

1 Wireless Course in 20 Lessons.

1 ABE10000 Fixed Variable Condenser.

1 FK10010 Junior Fixed Condenser.

1 No. BEK9240 Rotary Variable Condenser.

1 No. AK2501 Code Chart.

1 Blue Print of Connections.

Wire for Connections.

No. ADEK3030 "Electro" 300 Mile Receiving Outfit, complete as **$14.50** described ....

Shipping weight 12 lbs.

"Electro" 500-Mile Receiving Outfit

An outfit which will give results comparing very favorably to those given by most so-called "professional" sets. Tunes perfectly and has perfect interference protection. The range is very conservatively stated and is based on the use of a 4 wire L aerial 100 ft. long and 40 ft. high.

Outfit consists of:

1 No. HK14000 Professional Loose Coupler.

1 No. AEK7777 Universal Detector.

1 No. BEK8487 Loading Coil.

1 pr. No. DX8070, 2,000 ohm Receivers.

1 Complete Wireless Course in 20 Lessons.

1 No. BEK9240 Rotary Variable Condenser.

1 No. FK10010 Junior Fixed Condenser.

1 No. AK2501 Code Chart.

1 Blue Print of Connections.

Wire for Connections.

No. AHEK3050 "Electro" 500 Mile Receiving Outfit, complete **$18.50** as described ....

Shipping weight 25 lbs.

"Electro" 1000-Mile Receiving Outfit

The outfit you have been looking for. Simple to operate, sure to give results and beautifully finished. Its range of 1,000 to 1,500 miles is certain with an L aerial of 4 wires, 150 ft. long and 60 ft. high. Will receive undamped wave stations with a tikker and has frequently received foreign messages in New York. Has remarkable tuning capacity and perfect interference prevention.

Outfit consists of:

1 No. HX14000 Professional Loose Coupler.
1 No. HEK4500 Loading Inductance.
2 No. DX9241 Rotary Variable Condensers.
1 pr. No. FX1305 2,000 ohm Trans-Atlantic Phones.

1 No. CEK8888 Detector.
1 Blue Print of Connections.
1 Complete Wireless Course in 20 Lessons.
1 No. AK2501 Code Chart.
Wire for Connections.

No. CCEK3100 "Electro" 1,000 Mile Receiving Outfit, as described **$33.50**
Shipping weight 30 lbs.

"Electro" 1500-Mile Receiving Outfit

Rated at 1,500 miles but has done 3,000 miles very regularly on a 6 wire aerial 175 ft. long and 80 ft. high. The finest example of an assembled instrument outfit possible to produce. The tuning range is unusual, its work on undamped waves with a tikker perfect and its every part as substantial as it is possible to make it. The finish on every instrument is superb.

Outfit consists of:

1 No. AFX1399 Navy Type Loose Coupler.
1 No. HEK4500 Loading Inductance.
2 No. DX9241 Rotary Variable Condensers.
1 pr. No. GX6666 3,000 ohm "Government" Receivers.

1 No. CEK8888 Detector.
1 No. ABE10000 Rotary Variable Condenser.
1 Complete Wireless Course in 20 Lessons.
1 No. AK2501 Code Chart.
Wire for Connections.

No. DBGE3015 "Electro" 1,500 Mile Receiving Outfit, as described **$42.75**
Shipping weight 35 lbs.

"Electro" 3-Mile Sending Outfit

A simple untuned combination of instruments that will easily send from 1 to 3 miles with a 50 ft. 2 wire aerial 30 ft. high.

Outfit consists of:

*1 No. CCK1087 ½-inch Bull-Dog Spark Coil.

1 No. IE9221 1 pint Leyden Jar.

1 No. EK9220 Spark Gap.

1 No. CE1118 Telegraph Key.

3 No. CK1001 Dry Cells.

1 No. AK2501 Code Chart.

1 Complete Wireless Course in 20 Lessons.

1 Blue Print of Connections.

Wire for Connections.

No. FX2003 "Electro" 3 Mile Sending Outfit, as described...... **$6.00**
Shipping weight 12 lbs.

* This outfit can also be had for a range of 5 to 8 miles by substituting one of our No. DGE1088 one inch Bull-Dog Spark Coils and adding 3 No. CK1001 Dry Cells, making up our

No. HBE2008 "Electro" 8 Mile Sending Outfit, as described.... **$8.25**
Shipping weight 15 lbs.

"Electro" 15-Mile Sending Outfit

An outfit that any amateur can be proud to own. Its range is conservatively stated and has been greatly exceeded on a 75 ft. 4 wire aerial 40 ft. high. Its tuning is sharp and the emitted wave very penetrating. Finish excellent throughout.

Outfit consists of:

1 No. HX1089 2-inch Bull-Dog Spark Coil.

1 No. DX8271 Electro Helix.

1 No. EK9220 Spark Gap.

1 No. BEK9260 Condenser.

1 No. ABE1117 Telegraph Key.

8 No. CK1001 Electro Dry Cells.

1 Blue Print of Connections.

1 Complete Wireless Course in 20 Lessons.

1 No. AK2501 Code Chart.

Wire for Connections.

No. AHX2015 "Electro" 15 Mile Sending Outfit, as described.... **$18.00**
Shipping weight 25 lbs.

"Electro" 50-Mile Sending Outfit
For 110 Volts A. C. or D. C. Current Only

A reasonably priced outfit that has done wonderful work. A range of 75 to 100 miles is not at all unusual. Gives a high pitched and sharp note that is easily read and very dependable. The tuning is very simple and unusually sharp.

Outfit consists of:

1 No. GGE8050 ½ K.W. Transformer Coil.

1 No. BHE8000 Electrolytic Interrupter.

1 No. DX8271 Electro Sending Helix.

1 No. BX9212 Wireless Key.

1 No. AEK10099 Key Condenser (to absorb spark).

1 No. DGE531 Glass Plate Condenser.

1 No. DE8001 Fused Knife Switch.

1 No. AKX2382 Rotary Spark Gap.

1 Blue Print of Connections.

1 Complete Wireless Course in 20 Lessons.

1 No. AK2501 Code Chart.

Wire for Connections

No. CBEK2050 "Electro" 50 Mile Sending Outfit, as described.. **$32.50**
Shipping weight 90 lbs.

"Electro" 100-Mile Sending Outfit
For Operation on 110 Volts Alternating Current Only

A commercial outfit for the price of an amateur one. Its high grade apparatus as combined here has sent as far as 207 miles using a 4 wire aerial 100 ft. long and 60 ft. high. Tunes sharply and clearly and holds its adjustment. It is as dependable as any professional outfit can be made.

Outfit consists of:

*1 No. CAX9281 ½ K.W. Closed Core Transformer.

1 No. BX9212 Wireless Key.

1 No. AEK10099 Key Condenser (to absorb spark).

1 No. DGE531A Special Glass Plate Condenser.

1 No. FX9600 Oscillation Transformer.

1 No. AKX2382 A.C. Rotary Spark Gap.

1 Complete Wireless Course in 20 Lessons.

1 No. AK2501 Code Chart.

Wire for Connections

No. DEX2100 "Electro" 100 Mile Sending Outfit, as described.. **$55.00**
Shipping weight 125 lbs.

* The same outfit can be supplied with a 1 K.W. Transformer and Condensers giving it a range of over 200 miles at our extra price of $72.00, making the entire

No. AAGX2200 "Electro" 200 Mile Sending Outfit, as described.. **$117.00**
Shipping weight 225 lbs.

RADIOCITE

TESTED FOR SENSITIVITY

The most wonderful of all Wireless Crystals

Use Radiocite in your detector and then forget it

The Wireless Crystal that DON'T Jar Out

RADIOCITE is the most wonderful of all radio crystals. It is more sensitive than Galena and far more sensitive than ANY other crystal or mineral. RADIOCITE is a specially selected grade of a rare crystal chemically treated by our own secret process.

The mineral that looks like liquid gold. It has a highly, wonderfully polished surface giving it a perfectly burnished appearance. This crystal is now in use by several governments, and is conceded to be the most satisfactory of all. It is used with a medium stiff phosphor bronze spring, or with a stiff silver wire, about No. 30 B. & S. Gauge. One of the important features of RADIOCITE is that it does not jar out easily. Each crystal is **tested out individually for sensitivity** and guaranteed. RADIOCITE comes packed separately in a box, wrapped in tin-foil. Full directions for use accompany it. RADIOCITE can be mounted like any other crystal; it may be clamped between springs, but it is best to set it in **Hugonium** soft metal. Money refunded if our claims are not substantiated.

Shipping weight 2 oz.

No. EK3939 Generous piece of tested RADIOCITE. Prepaid **$0.50**

The one up-to-date mineral which every amateur must have.

Cleveland, Ohio,

Electro Importing Co.
    233 Fulton St.,
    New York.

Gentlemen :—

Your piece of radiocite received in excellent condition and am glad to inform you that it is without doubt the best mineral ever put on the market. It has any silicon or galena beat forty different ways and back again. I have tried it out on an indoor set consisting of a piece of bare copper wire 20 feet long, a gas pipe ground, a forty cent detector and a pair of 2000 Ohm phones. This set was used merely for the purpose of testing Radiocite and the results obtained "knocked me off my feet." I have not yet tried it on my big set but if it works as good as it did on the small set—why, I'll have "some" set.
    Yours truly,
        L. PLACEK,
        316 W. 84th St.,
        Cleveland, O.

The "Electro" Radioson Detector
"THE ULTRA SENSITIVE ELECTROLYTIC"

(Patents Pending.)

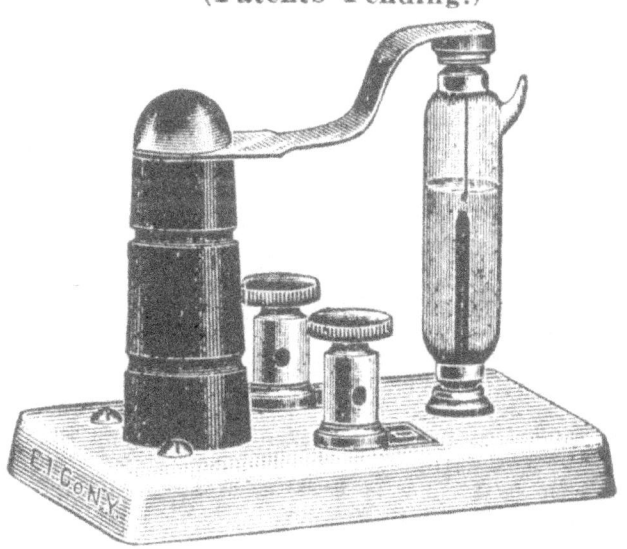

NO. DEK9300

This detector to-day is without question one of the most sensitive, and from an operating standpoint, the most satisfactory one manufactured. The Radioson is far more sensitive than most detectors and will bring in messages which cannot be heard with other detectors.

The Radioson Detector is the outcome of years of expensive experiments and embodies features new and unique.

The Radioson Detector is the only detector so far developed which needs no adjusting and cannot get out of adjustment. It cannot be knocked out by nearby sending stations, never loses its sensitivity and messages come in clear and distinct **even while the detector is shaken violently.** You no longer lose part of an important message because your detector lost its adjustment if you use the Radioson Detector.

The acid as well as all other parts are sealed in the detector. It is absolutely clean and safe, and it is adjusted to its highest sensitivity at the factory. Every Radioson cartridge undergoes five different tests before it is finally sealed. You cannot change the adjustment without smashing the glass or by passing a high tension current through it. The Radioson detector is always ready without adjustment on your part.

The Radioson is clean and compact and easy to handle. It works as well on a shaky table as on a concrete foundation. For receiving on board an aeroplane, shaky boat, train or automobiles where violent vibration is inevitable, the Radioson cannot be matched and this regardless of its higher sensitivity over other detectors.

It is necessary to use two dry cells (three volts) in connection with the detector. These cells may be of very small size, such as a flashlight battery. A curious part of the improved Radioson is that it does not sound at all like an electrolytic detector, but the sound coming in over the telephone receivers is exactly the same as that of a crystal detector. The sound is much sharper and clearer than the ordinary electrolytic type.

The **RADIOSON** practically requires no attention, it is always ready for use and the operator never loses part of a message on account of bothersome as well as annoying adjustments common to **every other** detector.

Another curious feature of the improved Radioson is that it tends to become more sensitive with age; the wireless waves passing through it seem to have a beneficent effect upon it. It does not matter which way the Radioson is hooked up, that is, whether the anode or the cathode is connected to the aerial; it works with the same intensity either way. In this respect it is exactly like a crystal detector and as a matter of fact works exactly like one.

The Radioson can only be used with a 2,000-ohm headset or one with higher ohmage. Lower resistances than 2,000 ohms tend to shorten the life of the detector.

The Radioson is absolutely guaranteed by us in all respects. We guarantee safe delivery to you under all circumstances. We will furthermore refund your money to you upon proof that the Radioson is not exactly as represented by us.

Gentlemen:— Knox, Pa.

The **Radioson Detector** I received is a "Peach." I am using it alongside of a very sensitive Galena and it is not only louder, but clearer and always in adjustment. V. E. Smith.

The "Electro" Radioson Detector
(Continued)

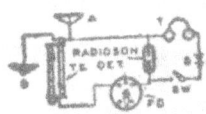

The Radioson is only sold complete as shown. Radioson exchange cartridges are only furnished to users of the instrument providing the original cartridge is returned to us either whole or broken.

Specifications: The Radioson consists of a heavy insulated base, on this is mounted a very large solid hard rubber standard, which supports the heavy nickel plated brass spring. The spring holds the Radioson cartridge in place by a positive string action.

The cartridge is easily snapped in and out by simply lifting the spring upwards. There are two extra large nickel binding posts. Size over all 4x2½x3½. All metal parts are triple nickel plated and highly polished. **Base is felt covered.** This extremely neat instrument has already been introduced in a number of commercial radio stations.

The very highest degree of sensitivity of the Radioson Detector is brought out by the use of our No. BX9255 Rotary Potentiometer. On account of the very minute amount of current used by the Radioson only a potentiometer will provide the fine control that will bring out the best that is in this remarkable detector. Of course, a potentiometer is not absolutely necessary but since you will want the best results attainable it will pay to use a potentiometer with your Radioson.

No. DEK9300 "Electro" Radioson Detector (complete)............ **$4.50**
Shipping weight 2 lbs.

No. BEK9301 "Electro" Radioson Cartridge only (see note above) **$2.50**
Shipping weight 1 lb.

RADIO "DE LUXE" CRYSTAL SET
FOR COMMERCIAL, NAVY AND ARMY OPERATORS

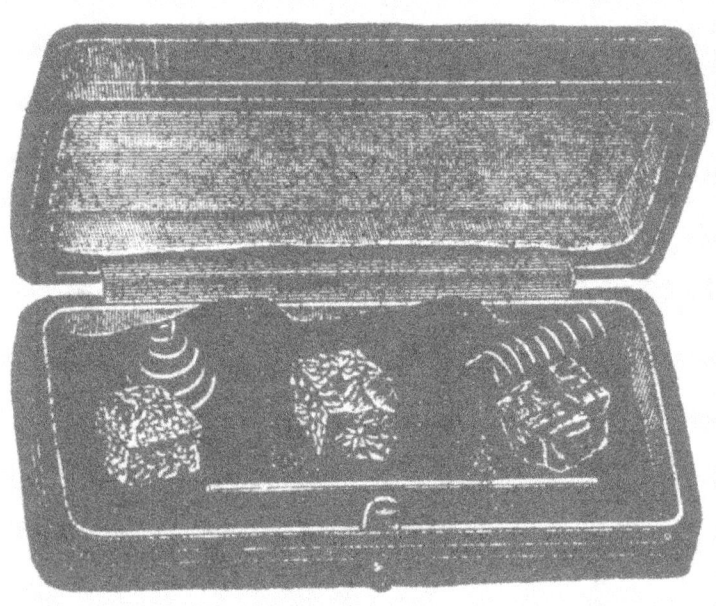

No. BX3131

Here is a very efficient outfit which every Radio Operator should have on his table or carry in his pocket. It is: **"The right thing in the right place, and the right place for the right thing."**

No more hunting around for that piece of crystal when the signals are not coming in strongly, no more soiled and broken crystals lying around in drawer's corner.

Our Radio "De Luxe" Crystal Set not only obviates this, but the high class minerals furnished with the outfit will prove a boon for every operator.

The outfit consists of a water- and dust-proof, air-tight box of special construction, as per illustration. The box can be carried easily in the pocket on account of its flat and neat shape.

It contains:
One piece of **tested** Radiocite.
One piece of **tested** Galena.
One piece of **tested** Silicon.

Furthermore, one **18 Karat Gold Cat-whisker** and two **phosphor bronze** cat-whiskers of different shapes.

It is not necessary for us to indulge here upon the merits of our **"Radiocite"** Crystal, as it is in universal use to-day in all well equipped Radio Stations; we will, however, add a few words of explanation concerning the quality of the two other minerals which we furnish with our set:

We use only **Galena** of the best and purest grade, especially selected cubic crystals, carefully **tested** and ultra sensitive.

Our Silicon is fused material, imported by us from England, and we have a good sized stock of same always on hand. Every piece is selected and **tested**, the same as our Galena and Radiocite.

Our tested Minerals should be handled only with pincers, never with bare fingers. We recommend strongly the use of the Gold Catwhisker with our "Radiocite." It is epecially invaluable on board ships, as the gold can't rust and no oxidation can set in between the point of the catwhisker and the mineral.

No. BX3131 Radio "De Luxe" Crystal Set, as described above................ **$2.00**
Shipping weight 1 lb.

THE "ELECTRO" RADIOCITE DETECTOR

NO. CEK8888

The "Electro" Radiocite Detector With Gold Catwhisker

$3 ⁵⁰

POSITIVELY THE MOST SENSITIVE CRYSTAL DETECTOR MADE

FEATURES

Gold Catwhisker	Ultra-Sensitive
Bakelite Base	¾ in. Felt Sub-base
Non jar-out	Adjustment Lock
Quadruple adjusting Range	Non Surface-leaking
Long distance tested	Rotary Detector Cup

There are to-day hundreds of crystal detectors on the market—many good ones, and more bad ones. However, but few of them are built along scientific lines, and most of them leave it to you to select whatever crystal you care to use.

Common faults with other detectors:

Wood or marble bases. Both leak, and badly at that, particularly the latter, due to metallic veins.

Nearly all detectors are easily "jarred" out.

Nearly all have catwhiskers or contact points which oxidize readily.

Most detectors are hard to adjust—and don't keep the adjustment, because they have no locking arrangement.

The "Electro" Radiocite Detector
(Continued)

They are only as sensitive as the crystal you use with them.

The "Electro" Radiocite Detector constructed along scientific lines by our Mr. H. Gernsback, is the outcome of ten years of experimenting to develop a REAL detector which should have none of the many faults listed above. Not alone has Mr. Gernsback succeeded in solving the baffling problem, but the Radiocite Detector has many other fine points not found in any other instrument. Mr. Gernsback in whose laboratory can be found almost every style of detector ever made to-day uses no other type than the Radiocite. You will feel the same way once you have used it for ten minutes.

The Radiocite Detector combines science plus good construction, plus highest sensitivity known to-day, plus horse-sense. No funny springs, levers, balls or other expensive freak contraptions are used. But what we do use is the best that money can buy. Everything is on a generous scale, no skimping is allowed.

CONSTRUCTION

BASE.—We use Bakelite ⅜ IN. THICK the best insulator known to-day, as well as the most expensive. Marble, Wood, Composition, Hard Rubber ALL LEAK, either due to poor inherent insulation or by **surface leakage.** On a damp day the best hard rubber base leaks, because it condenses moisture, thus making a shunt between binding posts, etc. Bakelite does not do this. Neither does it warp or crack. IT IS PERFECT.

ADJUSTMENT.—A heavy bronze casting, triple nickel plated WEIGHING ¼ lb. carries a very heavy shaft on both sides of which are attached two hard rubber knobs 1¼ in. in diameter. In the center of the shaft a small tubular rod is inserted, carrying the little thumbscrew "B," which in turn holds the catwhisker wire. By this simple method the catwhisker can be exchanged for another one in five seconds. Now look at the illustration and observe the slot in the large casting.

This simple slot makes it possible to LOCK THE CATWHISKER by means of the hard rubber adjusting knob "A" WITHOUT DISTURBING THE CATWHISKER ADJUSTMENT IN ANY WAY WHATSOEVER. The knob "A" simply locks the central rotating shaft, **without displacing the catwhisker as much as a millionth of an inch.** No other detector has any such wonderful locking feature.

Two knobs are used for the following reason: For very fine adjustment one hand rests on one knob while the fingers of the other grasp the other knob. The hand resting on the one knob simply acts AS A BRAKE. A moment's reflection will show that this method is infinitely better than a single hand adjustment.

Moreover with this detector, BOTH HANDS REST ON THE TABLE while adjusting. All these features must appeal to the wireless enthusiasts who know how difficult it is to adjust the average detector.

CATWHISKER.—We furnish two of these. One of 14 KARAT GOLD, impossible of oxidation, the other of phosphor bronze.

Most detectors are out of adjustment a large part of the time, simply because the tip of the fine catwhisker oxidizes, i.e., rusts. Gold does not oxidize hence we use it. You will be surprised how little adjusting the Radiocite requires.

QUADRUPLE ADJUSTING RANGE.—This wonderful range is only to be found in the Radiocite Detector. It positively beats everything for quick and complete searching out of the most sensitive crystal spot.

1. Rotating the two large knobs, adjusts the catwhisker for best contacting pressure.

2. Pushing the knobs from one side to the other (⅜ in. movement allows for this) gives the catwhisker ample lateral motion.

3. The Rotating Detector Cup, rotated by means of the knurled insulating ring, serves to bring almost every point of the crystal under the catwhisker.

4. Sliding the Detector cup backwards or forwards completes any possible adjusting that can be imagined.

Finally the detector can be screwed on the table with the crystal towards you or away from you. The adjusting is accomplished equally well either way.

NOT JARRED OUT.—This is a very important feature. The long fine catwhisker wire is so light that it needs a very heavy knock to displace it. To deaden any jar or knock we employ a ⅜ IN. THICK SOFT FELT SUB-BASE (not shown in illustration). This makes the detector practically jar proof. Think of this when you want to buy a detector.

RADIOCITE CRYSTAL THE MAIN FEATURE.—"A detector is no better than its crystal." We come out with this strong claim, supported by evidence from thousands of users:

Radiocite is the most sensitive crystal known to-day barring none. It is more sensitive than galena, zincite, or silicon. Evidence?

We have thousands of unsolicited testimonials on hand. We could print pages and pages of them, but lack of room forbids. If you want to know more about Radiocite refer to the page where Radiocite is listed in this catalog.

EACH DETECTOR TESTED FOR SENSITIVITY AND FOR LONG DISTANCE. Before the Radiocite crystal is set in the Rotary cup with "Hugonium," it is tested by two operators, under actual working conditions. Only the very best and most sensitive crystals are used in this detector. You will be amazed at the sensitivity of this crystal; there is nothing like it. Long distance records are broken every week with Radiocite, 1,000, 2,000 miles are every-day performances. Will you use the best?

All workmanship and finish highest throughout. All metal parts are triple nickel-plated and hand-buffed. Two very large binding posts are used. The Bakelite base is highly finished. The bright nickel on the black Bakelite base gives the whole instrument a rich appearance, not possible to reproduce adequately in our illustration, which at best does the instrument scant justice.

Size 3x4½x2¾ in. Shipping weight 3 lbs.
No. CEK8888 "Electro" Radiocite Detector, as described....... **$3.50**

"Electro" Cardboard Tubes

We have had such a persistent demand for cardboard tubes for winding radio coils, Tesla coils, high frequency coils, phantom coils, and for all other kinds of similar experiments, that we finally decided to list these.

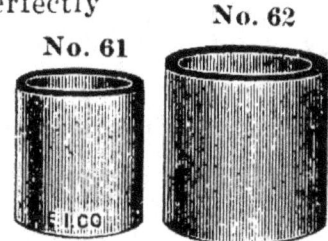
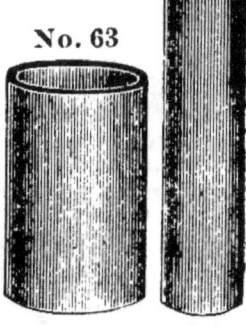

All our tubes are made of special gray seamless board and there are no seams showing on any part of the outside of the tube. They are perfectly smooth and clean, while the most important part is that all our tubes are absolutely accurate and perfectly round. These tubes are built of specially treated paper, are aged and are guaranteed not to shrink even after being treated with shellac or other compounds after winding on the wire. This is an important consideration, as most ordinary mailing tubes are not fit for winding purposes because they shrink after a certain time, the wire then becoming loose. This is absolutely obviated by our specially treated tubes, and they will save quite a good deal of annoyance, time and material wasted as must become apparent to any experienced Experimenter who has ever had to do work of this kind. Yes, our prices are high, but our tubes are worth it.

The color of the tubes is light gray; sizes over all are as follows:

No. 61—4-11/16 in. high; 4¼ in. diameter; ¼ in. wall
No. 62—5⅜ in. high; 5¼ in. diameter; ¼ in. wall
No. 63—7¼ in. high; 4⅝ in. diameter; ⅛ in. wall
No. 64—12 in. high; 2¾ in. diameter; ¼ in. wall
No. 65—30 in. high; 5¼ in diameter; ¼ in. wall

The No. 65 tube is not shown in our illustration as it is a very large size. This type is used in our regular No. 4500, 15,000 meter loading coil. We guarantee these tubes will please you. Once used, always used.

At the present time we are not prepared to furnish any other tubes than the ones listed here, although we will take orders of special sizes for quantities. In that case it must not be for less than 300 tubes. Smaller orders cannot be filled. Get our prices first.

No. 61 "Electro" Cardboard Tube, as described, each............ **$0.25**
Shipping weight 1 lb.

No. 62 "Electro" Cardboard Tube, as described, each............ **$0.35**
Shipping weight 1 lb.

No. 63 "Electro" Cardboard Tube, as described, each............ **$0.35**
Shipping weight 1 lb.

No. 64 "Electro" Cardboard Tube, as described, each............ **$0.25**
Shipping weight 1 lb.

No. 65 "Electro" Cardboard Tube, as described, each............ **$0.75**
Shipping weight 5 lbs.

The "Electro" Galena Detector

WITH A PIECE OF TESTED GALENA AND ROTARY DETECTOR CUP

NO. ABE9700

The preference of many amateurs for a light contact crystal detector has caused the advent of our Galena Detector. To evolve a detector of this type demanded no particular skill, but to construct a detector with **EVERY** advantage heretofore enjoyed by this class of detectors required much and long study. And above all this detector is presented at a lower price than many more costly detectors and far inferior to ours.

With this detector it isn't a case of simply another detector on the market with the usual failings of those now on the market. It's the application of a really new method of adjusting a sensitive cat-whisker so it will apply as light or heavy a contact as is desired and all this by use of the simplest means imaginable, yet so simple that it wasn't thought of all these years. The surest sign that the "Electro" Galena Detector is a hit is the compliment that we have been paid by a competitor who copied it exactly in appearance though not in material or working quality. Better make sure you get the best by ordering the "Electro" to-day.

The base is of solid hard rubber composition, ¼ in. thick, solid standard heavily nickeled and polished adjusting screw of hard rubber composition 1 in. in diameter, contact spring of phosphor bronze, nickel plated and polished; crystal contact of phosphor bronze wire properly coiled and pointed. Binding Posts are our Standard Hard Rubber. The cup is surrounded by a knurled fibre ring, and can be rotated, so that every point of the Galena can be reached.

By reason of the fact that the contact spring moves through an arc and the crystal cup moves on an eccentric, every spot on the cup can be touched. The spring being held down by a screw cannot slip and lose its adjustment due to vibration. Size of Detector, 3½x3½ in.

No. ABE9700 Detector complete as described. **$1.25**
Shipping weight 1 lb.

No. AEK9701 Detector as described but with a TESTED "RADIO- **$1.50**
CITE" crystal, the best there is......................
Shipping weight 1 lb.

A piece of TESTED GALENA, set in the Detector-cup with HUGONIUM soft metal is furnished with No. ABE9700 Detector and a piece of TESTED RADIOCITE set in the Detector-cup with HUGONIUM soft metal is furnished with No. AEK 9701 Detector.

The "Electro" Universal Detector Stand

WITH ROTARY SLIDING DETECTOR CUP

NO. AEK7777

Our Universal Mineral-Crystal Detector Stand was devised by us after long experimenting and stands in a class by itself. It is used chiefly for experimental purposes and has the most sensitive arrangement of any detector on the market. It is hardly necessary to waste words on the superiority of this instrument over other similar ones. By studying the illustration the many excellent features of this detector will appeal even to the layman.

The crowning achievement, however, is found in our new rotary sliding cup. The brass cup has a massive knurled fibre ring and the lower part of the cup is fashioned in such a manner that it can slide back and forward in the metal slide. ANY PART OF THE DETECTOR SUBSTANCE can thus be touched by the detector point. Ours is the first detector to achieve this. Furthermore, the entire cup can be slid out entirely and another one substituted in 5 seconds. The upper double spring arrangement has a blunt brass point, to make contact with the crystal or mineral. A phosphor bronze cat-wisker contact not shown in the illustration is supplied with each detector.

The novelty is that with this Detector we furnish a quantity of SOFT METAL which is packed around the crystal or mineral into the detector cup. This soft metal, HUGONIUM, is furnished in a small bottle and when placed in hot water immediately melts. Upon cooling it becomes hard as copper. All metal parts are nickel plated and polished, and mounted on an insulating base. Two hard rubber binding posts are provided.

When you buy a detector you are buying the most important part of your wireless receiving outfit. The finest receiving set is no better than its detector. When you buy the "Electro" Universal Detector stand you are buying a time tested piece of apparatus that works not sometimes or once in a while, but every time. You can't make a mistake by buying the best at the price of most lower grade articles.

THIS DETECTOR IS EQUIPPED WITH A TESTED GALENA CRYSTAL:

Size over all 2½x4½x3½ in.

No. AEK7777 New Universal Detector Stand, as described.. **$1.50**
Shipping weight 1 lb.

No. CE7778 Hugonium, soft metal, to mount crystals or minerals, oz. bottle. Shipping weight 4 oz........ **$0.35**

No. CE7778

Gentlemen:—   Jersey City, N. J.

Am just after receiving my Universal Detector and Buzzer and am very much pleased with both, especially the Detector.

ANDREW SCHMIDLAPP.

The "Electro" Tuner
3300 METERS

PATENTED FEB. 1, 1910

NO. DX8486

The "Electro Tuner" which we present is undoubtedly one of the best examples of a high grade commercial tuner. We were the very first firm building this style, and there is no doubt whatever that this tuner is the most popular one we' manufacture. The type described herewith is the tenth style evolved, and we believe it is impossible to further improve on it. Our long years of experience vouch for superority.

A Few Words to the Uninitiated.

Take two violin strings, having equal length and stretched with equal tension. If now one is sounded with the bow, the other will sound in unison, although it has not been touched. **The two strings are tuned,** or both have the same tune. The same thing is true of piano strings and tuning forks.

The tuning coil for wireless purposes accomplishes a similar purpose. Roughly speaking, if your friend has an aerial 60 feet long and you have an aerial 40 feet long, theoretically you cannot receive from his station because you are 20 feet short. Now the tuning coil, having hundreds of feet of wire wound on its cylinder, makes good your deficiency. By moving one of the sliders your aerial (connected to the tuner) will be made artificially longer. Finally the slider passes a point where the total length of the tuner wire has reached 20 feet. This, added to the aerial, gives 60 feet total—you are in tune with your friend's station. Thus a tuner accomplishes the purpose to tune in in other stations, to tune out unwanted ones, etc., etc. In other words, though your aerial is not very big, you can, with the aid of the tuner, select the signals from other stations having larger aerials than your own.

The wave length of the "Electro Tuner" with 100-foot flat top 4-wire aerial 100 feet high is 3300 meters. This tuner is perfectly adapted to receiving from undamped wave stations and if 3 or more of them are used in series on a fair sized aerial you can receive from the European stations perfectly. According to the latest researches it is BARE WIRE WOUND by our special process. The convolutions approach up to 1/100 inch and are wound with amazing precision, making finest tuning possible.

Gentlemen:—   Moscow, Idaho.
    I received order all O. K. The tuner was better than I had expected. The wire and ground clamp are also very good.   GEORGE CURTIS.

There are over 300 convolutions of copper wire, and this tuner will tune as accurately as ½ meter. The wire is secured by means of a secret process and we will exchange any coil WITHIN TWO YEARS IF THE WIRE SHOULD COME LOOSE OF ITS OWN ACCORD.

All woodwork is of best seasoned oak, hand rubbed finish cf the highest order. The base has holes so the instrument can be screwed to the table or wall. The two metal slider rods are double nickeled and highly polished. Our well-known hard rubber sliders are used, one BLACK for ground, one RED for aerial. This feature is found only on our instruments.

Our patented sliders are equipped with a medium hard BRONZE BALL which does not wear down the wire of the tuner. Our slider in this respect is the only one that can claim this truthfully. Besides being made of hard rubber, **ground through your body by way of the slider is impossible.** There are four of our famous hard rubber binding posts, which give the tuner that "classy" appearance. Each post is marked. W stands for the wire convolutions; S for slider.

Size of tuner over all, 13x7x7 inches.
Shipping weight 6 lbs.

No. DX8486 "Electro Tuner" double slide, as described.......... **$4.00**

The "Electro" Tuner, Jr.
PATENTED FEB. 1, 1910

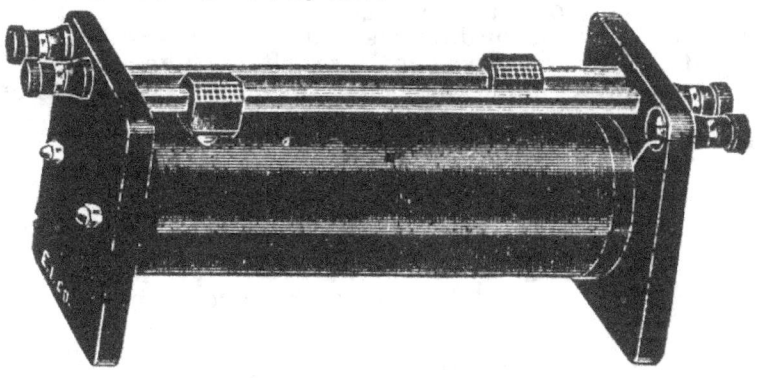

NO. BX9950

While the "Electro" Tuner previously described may be used to "tune in" for the largest stations, we have had a large demand for a Double Slide tuner of a smaller design and present our friends with our "Electro Tuner, Jr."

It is wound with about 300 turns of bare copper wire, and instead of one slider two are used. This gives a finer adjustment. Our Patented BALL BEARING Slider is used on this coil. It has 4 binding posts.

The construction of this coil is the acme of perfection. Coil ends are of molded hard rubber giving the instrument a beautiful appearance. All metal parts finished in nickel plate.

The Rolling ball touches only one wire at a time and can not possibly wear out the wire. No other coil can claim this. Springs rubbing against the bare wire, as found in other coils, wear the wire down quickly, and the coil must be rewound. This is impossible with our coil. Ours is cheaper at the start, and cheaper in the end. Here's how we make this little marvel. ENDS—polished hard rubber composition. A fine insulator wonderfully strong and beautiful to look at.

TUBE—Non-seamless—carefully dipped and finished. They are wound with bare wire (the best way). The wire can't come loose and will be replaced if it comes loose without abuse within two years.

SLIDER RODS—Solid Square rods highly nickel plated and polished. They are made to last and they do. On the rods are our wonderful patent sliders, one red and one black—an exclusive feature found only on our goods.

BINDING POSTS—are solid hard rubber composition. Can't short circuit, make perfect contact and look rich.

Sizes, 8 inches long, 3¾ inches high, 3¼ inches wide.

No. BX9950 "Electro Tuner, Jr." (double slide) as described.... **$2.00**
Shipping weight 2 lbs.

The "Electro" Loose Coupler
RECEIVING TUNING TRANSFORMER
Patented Feb. 1, 1910

NO. DBE12002

While an ordinary tuning coil is admirably suited for ordinary work it is not a success where exceedingly fine tuning is required. In fact, even the best tuner cannot tune within 10 per cent. accuracy. Furthermore, now that so very many stations are working simultaneously, we must have an instrument which is capable of tuning to an exceedingly fine degree and be able to ABSOLUTELY tune out ANY unwanted station.

We experimented for months before we produced a loose coupler within the reach of everybody. Not alone did we succeed but we improved the old types to such an extent that ours has a far greater selectivity than any similar instrument on the market NO MATTER WHAT ITS PRICE. Certain far off stations come in quite loudly even if the secondary is pulled clear out as far as it will go, that is, the air distance between primary and secondary is fully 2 inches. We found the connections as per diagram to give best results. The variable condenser is especially recommended and will be of considerable value. Any detector can be used, of course. Personally we prefer the Radioson Detector as the signals come in very much louder.

The construction of the "Electro" Loose Coupler is of the highest perfection.

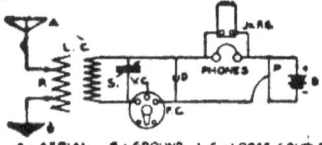

A - AERIAL   G - GROUND LC - LOOSE COUPLER
D - ELECTROLYTIC DETECTOR FC - FIXED CONDENSER
P - POTENTIOMETER   B - BATTERY VC - VAR COND

Wood parts are of polished hard wood; metal nickel plated. The wire on the primary is bare wire wound after the latest process, ensuring high efficiency; 3 hard rubber binding posts and two generous metal posts are provided as shown. If the variable condenser is not used, post No. 1 remains unconnected.

The secondary is machine wound with green silk covered wire, as it would be quite impossible to wind the very fine wire otherwise. It is, of course, highly important that no wire of one layer should cover any other; in other words, the winding must be done with highest precision only made possible with a special winding machine.

The secondary, projecting from the right has a large hard rubber switch handle, which carries a nickel switch blade. This blade plays over 6 contact points, to vary the inductance. The secondary coil heads **ARE OF HARD RUBBER COMPOSITION**, the secondary slides **freely** on two beautifully nickel-plated brass rods. On the primary one of our patent sliders is provided as used on our other instruments. The secondary can be moved back and forth with the greatest possible ease and will not stick, or require two hands to move as is the case with even expensive makes. Our loose coupler is built to pick up wave lengths up to 800 meters and as the majority of commercial and government stations have only a wave length up to 600 meters, our instrument will be found to respond in practically all cases.

Adjustment: When connections are made and detector is adjusted, move secondary up to the centre of primary, then adjust slider till signals come in loud; then move secondary back and forth, while moving the switch knob back and forth, till position is found where signals are loudest. Now the variable condenser is adjusted. Dimensions: Length of base 12 inches, width 6 inches, height over all 6½ inches.

No. DBE12002 "Electro" Loose Coupler, as described.............. **$4.25**
Shipping weight 5 lbs.

The "Electro" Professional Loose Coupler

(RECEIVING TRANSFORMER)

The loose coupler which we present herewith is the outcome of long and careful experimentation to produce an article that really does away with all the objections of the ordinary loose coupler.

Our professional loose coupler has been carefully balanced and the secondary and primary have been wound according to the latest researches in this art. The diameter as well as the amount and the size of the wire is highly important and the type which we present herewith is unusually effective and we guarantee it to do anything and everything, even the most expensive loose coupler on the market to-day will do.

NO. HX14000

This coupler is made of **hand rubbed, piano finished mahogany throughout.** Primary winding is of bare copper wire wound by our special process and there is one of our well known patented Hard Rubber Ball Sliders conveniently located on the side. This slider makes perfect contact on only one turn of wire at a time and never wears out the wire. The secondary wound with green silk covered copper wire is calculated for long wave lengths and the crowning feature of it is the secondary switching arrangement attached to the secondary. There are 8 switch points to the rotary switch which is directly attached to the secondary. By means of its knob the secondary can be moved backwards and forwards and this arrangement gives the maximum of efficiency in the minimum of time, particularly when quick tuning is necessary. Thus the switch knob is used for switching in more or less secondary turns and for moving the secondary backward and forward at the same time.

This feature as a rule is only found in "Navy" style couplers and this is the first loose coupler ever placed before the public, making use of this expensive as well as ultra-efficient feature.

Kindly note that the entire secondary rotary switching arrangement is built of solid MOULDED HARD RUBBER, not wood or composition. Also note particularly that the secondary coil heads are of MOULDED HARD RUBBER COMPOSITION not wood.

Dear Sirs:—   Newark, Ohio.
I have one of your Loose Couplers, Fixed Condensers, Detector and 2000-ohm phones and am able to pick up Duluth, Minn. (DM), along with other stations. I can bring them in very plain with your loose coupler.
                                                CARL HOWARD.

The "Electro" Professional Loose Coupler
(Continued)

With this instrument, in connection with other good receiving apparatus, nearly all large stations can be heard without much trouble.

(See article by Mr. S. Curtis of the U. S. Navy in December, 1916, issue of THE ELECTRICAL EXPERIMENTER. Using this identical coupler in connection with other apparatus, Mr. Curtis on the U. S. S. New Jersey, laying off Massachusetts could hear the Nauen (Germany) station in broad daylight—8,000 miles!)

What other $8.00 Loose Couper could perform such an extraordinary record?

Only first class material is used in connection with this fine instrument. All nickel-plated work is hand buffed, not merely polished. You will be proud indeed to own this instrument.

There are five large nickel binding posts, two for the primary winding, one for the primary slider and two for secondary winding. The loose coupler has a wave length of 3,000 meters without the use of a loading coil.

The "Electro" Professional Loose Coupler is guaranteed to do the work of any professional loose coupler, regardless of its price. Note size of this loose coupler, base 15½x7¼x7½ high. Length of Primary is 5¼ inches; length of Secondary is 4¾ inches.

No. HX14000 Professional Loose Coupler **$8.00**
Shipping weight 10 lbs.

The "Electro" Navy Type "3 in 1" Coupler
PEER OF THEM ALL

NO. AFX1399

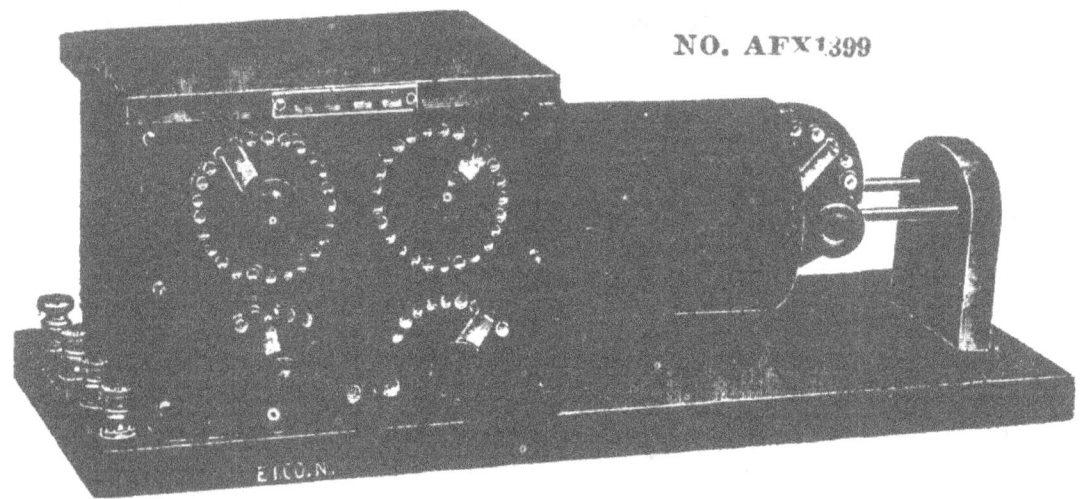

With this we present an instrument to the advanced Radio enthusiast that has not a counterpart on the market to-day. There are many Navy Type Couplers on the market now, but we feel confident that you will find in this instrument features that you never thought possible in such a Coupler. We claim it to be the acme of perfection of an instrument of this kind, no expense or money having been spared to make it such.

We call this Coupler "3 in 1" for the reason that it not only has all the Navy Type Coupler features, but in addition, with the same instrument, we furnish a high grade loading coil, as well as a variable condenser, the three instruments being all built in the one case. On the Navy Type Coupler no sliders are used whatsoever, the tuning being accomplished entirely by means of rotary knobs or switches. At the upper left side you will find twenty-four switch points for the primary tuning, while at the upper right there are twenty-four switch points for cutting in single turns, on the primary, thereby giving one turn to the coil for every switch point.

This latter feature is highly important when working with an Audion or a valve type Detector where exceedingly fine tuning is necessary; this improvement is one only found in professional couplers.

The third switch at the lower left side controls the variable condenser which is also necessary for extremely fine tuning, while at the lower right side is found the loading coil, by which long waves can be tuned in without taking recourse to a separate instrument.

The secondary, wound with silk wire, is calculated for extra long wave lengths. There are 8 switch points to the rotary switch, which is directly attached to the secondary. By means of its knob the secondary can be moved backwards and forwards and this arrangement gives the maximum of efficiency in the minimum of time, particularly when quick tuning is necessary. Thus the switch knob is used for switching in more or less secondary turns and for moving the secondary backward and forward all at the same time.

With this Coupler most all of the large stations can be heard with a fair sized aerial on account of its long wave length, and there will be but few stations indeed from which you cannot receive with this Coupler.

Only first-class material is used in connection with this instrument. All nickel plated work is hand buffed, not merely polished. There are four large binding posts at the left, two small binding posts for the telephone receivers in front. The secondary coil ends are of hard rubber composition. An absolutely distinct feature of this Coupler is found in the fact that the front part carrying the various switches is not of wood nor hard rubber, but is of **BAKELITE**, the latest electric product and more expensive than hard rubber; it also gives the highest electrical insulation to-day for this work. It is a black substance, almost unbreakable and offers much better insulation than hard rubber.

This Bakelite plate is placed at a slight angle in respects to the apparatus; this makes the working very much easier than if it was placed at right angles to the base.

The woodwork used throughout is mahogany, handrubbed, piano finish. You will be proud of this masterpiece of Radio instruments.

Dimensions are as follows: 19 in. long, 7¾ in. wide, 6¾ in. high. Shipping weight 14 lbs.

No. AFX1399 The "Electro" Navy Type "3 in 1" Coupler, as described. Price **$16.00**

The "Electro" Vario Selective Coupler

CABINET TYPE

In presenting this outfit the only introduction necessary, is consideration of the fact, that an outfit of this type convinced the **D. L. & W. Railroad** that long distance communication with trains in motion carrying small low aerials was not only possible, but eminently practical.

It consists of a highly selective induction coupler of the cabinet type in which all tuning is done by switches acting on switch points, eliminating sliders entirely. There are three of these switches: one marked primary, having 25 contact points, another secondary with 7 contact points, and the third a loading circuit for long

NO. FEK11000

wave lengths, having 7 contact points. This outfit can tune to wave lengths from 100 meters to 3,000 meters and its selectivity is so perfect, that with 4 stations sending at one time, we have been able to select any one station, eliminating the others entirely.

The finish is superb, the entire cabinet being made of highly polished mahogany, with switches controlled by hard rubber handles, and the binding posts and metal parts of brass, nickel plated. The size 9x9x2 in. and weight of this outfit (less phones and detector), being only 2 lbs., especially recommends it for service under conditions where space is at a premium or where weight must be kept down.

The loudness of signals received is due to the variometer effect introduced in this outfit which eliminates all open or dead ends in the windings.

This is one of the smallest and most compact long distance wireless receiving outfit manufactured and we particularly recommend it **FOR RECEIVING TIME SIGNALS** as sent out by the various U. S. Government Wireless Stations. The outfit may be used with any type of detector and any phones but we particularly recommend the use of the No. DEK9300 Radioson Detector and our No. GX6666—3000 ohm Government Phones.

No. FEK11000 "Electro" Vario Selective Coupler (no phones or
        detector). Shipping weight 5 lbs.................. **$6.50**
No. AHX11001 "Electro" Vario Selective Coupler, complete with No. GX6666
        Government 3000 ohm Phones and No. DEK9300
        Radioson Detector. Shipping weight 10 lbs........ **$18.00**

The "Electro" Loading Coil

In order to receive messages from stations using very long wave lengths it becomes necessary to use a loading coil in order to increase the natural wave length of the ordinary tuning coil or loose coupler. Our loading coil has a wave length of approximately 5,000 meters. If placed in series with either our No. DX8486 or No. BX9950 tuning coils, or our No. DBE12002 coupler (in series with the primary) or our "Interstate" outfit it becomes possible to catch time signals from the Arlington Government station using 2,500 meters wave length. These stations can not be heard with the ordinary tuner or coupler, as these instruments usually do not go beyond 600 or 800 meters wave length. The

NO. BEK8487

use of our loading coil enables one to receive messages from almost any station, no matter what its wave length, up to 5,000 meters, the capacity of your outfit is increased enormously as you can practically receive most any message from any radius.

There are six steps, each switch point representing approximately 800 meters wave length, and by simply revolving the knob most any wave length can be. obtained. Of course, it must be understood that either a loose coupler or a tuner must be used in conjunction with this instrument **as it cannot be used by itself alone.**

These progressive times when your amateur is only satisfied with receiving from only the most distant stations our "Electro" Loading Coil is of especial value for European stations use long wave lengths exclusively which cannot be received without a loading coil such as we offer here for so low and reasonable a price.

We cannot praise this wonderful instrument too highly and once used you will not do without it. It is made entirely of hard rubber composition with large hard rubber handle and hard rubber binding posts. All metal parts are nickel plated and highly polished; its size is 4 in. in diameter and 1½ in. in height. The diameter of the hard rubber thumb handle is 1 in. WE GUARANTEE SATISFACTION.

No. BEK8487 Electro Loading Coil, as described. Price........ **$2.50**

Shipping weight 1 lb.

The "Electro" "Trans-Oceanic" Undamped Wave Loading Coil

15,000 METER COIL

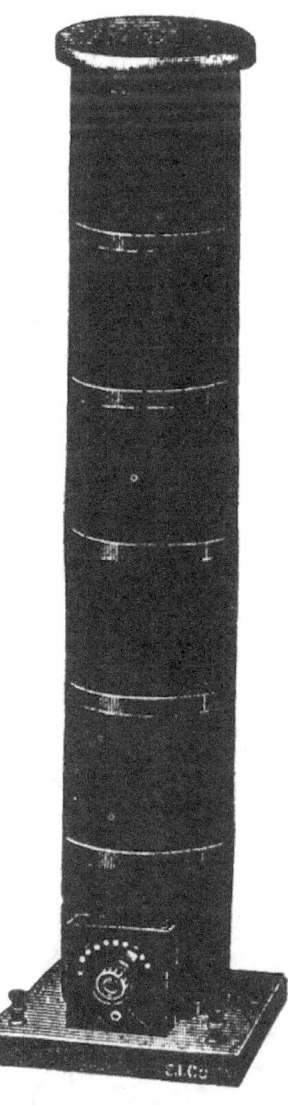

52 IN. HIGH

There is a distinct need of an extra long wave loading inductance for use in receiving the long distance undamped signals from such stations as: Nauen-POZ; Eilvise—OUI; Arlington—NAA; Tuckerton—WGG; Darien—NBA; Clifden—5CN; Sayville —WSL and over twenty other high powered, long wave stations in all parts of the world. As an example, the Nauen, Germany, station (call POZ) transmits on either of three waves, viz.,—6,300, 9,400 while 12,600 meters and 10,000 to 14,000 meter wave lengths are quite common among the newer, long range stations. What thrills and what fascination for the amateur, now that we have placed him in a position where he can receive messages daily from Germany on a moderately small aerial!

Such circuits as the Armstrong, which utilize vacuum tube oscillating relays, in order to hear stations working on waves of 10,000 to 15,000 meters length, absolutely require a first-class inductance such as we present to our patrons herewith. These waves can not be received otherwise.

We are confident that our "Trans-Oceanic" long wave tuning inductance will meet every demand that can be made of it. It is adjustable by means of a seven-point switch mounted on the base in a convenient position, as shown in the illustration, which by the way does the instrument but scant justice. The inductance of the winding has been carefully balanced and properly divided up in six equal sections, in the latest approved manner so that when used with a four wire, 300 foot flat-top, "L" shaped aerial, placed 100 feet above the ground, in conjunction with our Navy type or other large size loose coupler, wave lengths up to 15,000 meters and more can be easily tuned in. This considers that the loading inductance is connected in series with the aerial and the loose coupler primary.

The long wave lengths mentioned are also within your range when the coil is used in the **secondary** coupler circuits, as for instance in the Armstrong hook-up, which requires two of these coils for the secondary or vacuum valve circuits (for exact connections see **The Electrical Experimenter**, page 632, for March, 1916; also page 488, January, 1916, issue and page 337, November, 1915, issue) and one for the aerial or primary circuit unless you intend using a special large size loose coupler capable of tuning in 10,000 to 15,000 meter waves directly, when only two are required. If this inductance is employed with a 500 foot flat top, "L" design aerial, placed 100 feet above the ground, IT WILL TUNE UP TO 20,000 METERS WAVE LENGTH, when used with any standard large size coupler, such as our Navy type.

Ordinarily and when a small loose coupler of 3,000 to 4,000 meters wave length capacity, such as our "Navy type," is used with a vacuum valve detector ("beat" producer) the following auxiliary apparatus is necessary besides three of the "Trans-Oceanic" inductances:

Three .001 M.F. No. DX9241 and two .0004 M.F.; No. BEK9240 variable condensers; one No. ABE10,000 .003 M.F. fixed variable condenser. It it presumed of course that you have or intend to procure the loose coupler, vacuum valve and phones, which latter should be of 2,000 to 4,000 ohms resistance—the higher the better.

NO. HEK4500
15,000 METERS

The "Electro" "Trans-Oceanic" Loading Coil
(Continued)

If a "tikker" is used for interpreting the undamped wave signals then the apparatus required includes two "Trans-oceanic" loading inductances, one for primary and one for secondary circuits; a "Tikker" across the stopping (fixed) condenser, No. ABE10,000 type; our "Navy type" coupler; two No. DX9241 .001 M.F. variable condensers and 'phones. No detector or vacuum valve is necessary. The Tikker should make about 200 interruptions per second. Dr. de Forest found that a crystal detector reduced the signal strength on long distance reception but one may be used for ordinary work with the Tikker. The sound or pitch of the received signal can be altered as desired by varying the capacity of the variable condensers in the Armstrong circuits; in "Tikker" circuits the pitch is variable by changing the speed of the Tikker interruptions. Good hook-ups and a discussion on "Tikker" circuits are given on page 632, March, 1916, **ELECTRICAL EXPERIMENTER**.

The "Trans-Oceanic" long wave inductance measures 32 INCHES HIGH by 8 inches square at the base. The extra heavy tube is machine wound with a single layer of single silk covered, pure copper magnet wire of ample size to keep the ohmic resistance down to a minimum. SMALL COILS WOUND WITH FINE WIRE PRESENT A HIGH OHMIC RESISTANCE AND CONSEQUENTLY A HIGH DAMPING EFFECT. THIS INDUCTANCE HAS THE LOWEST DAMPING FOR ITS SPECIFIC INDUCTANCE VALUE, OF ANY SIMILAR INSTRUMENT ON THE MARKET— BAR NONE. The taps from each section are brought down inside the instrument and through the hollow base to the multi-point switch shown in the illustration. No more reaching up in the air and tiring your arms while adjusting the inductance. All wood-work is hand-polished mahogany, piano finish. Metal parts heavily nickel plated. All parts substantial and well designed. Nothing to wear out or collapse at a critical moment. The "Trans-Oceanic" is really a commercial instrument in all respects: in design, workmanship, efficiency and appearance. Order it to-day and be convinced.

No. HEK4500 "Trans-Oceanic" Undamped wave loading coil.
Price **$8.50**
Size 8x8x32 in. over all. Shipping weight 15 lbs.

The "Electro" Sliders

No. CE2222

The accompanying cut (actual size) shows our hard rubber ball-bearing slider (patented Feb. 1, 1910). It is the acme of perfection and surpasses in efficiency, quality, accuracy, any slider ever placed on the market.

As it is non-metallic, it slides over the rod with astonishing ease. **NO MICROPHONIC CONTACTS** are possible with this slider, no jars in the telephone. The brass ball is pressed evenly on the tuning coil wire, while the phosphor bronze spring which makes contact with the rod, presses firmly on the ball, ensuring perfect contact at all times.

It can never stick, but responds at once, quickly and with astonishing ease. ALL OUR INSTRUMENTS ARE EQUIPPED WITH **OUR PATENT SLIDER, WITHOUT EXTRA CHARGE.**

NOVEL FEATURE. We have devised the same slider in **BRIGHT RED COMPOSITION**, for the Aerial slide, and therefore equip all our double slide tuners, couplers, etc., with one **black and one red slider**. This original feature is a distinctive departure. We lead as usual. Sizes over all ¾ x ⅞ x ½ inch. Our new slider fits any ¼-inch square rod.

No. CE2222 Hard Rubber Slider (black) complete with ball and spring **$0.35**
No. CE2222a Hard Rubber Slider (red) complete with ball and spring **$0.35**
Shipping weight 3 oz. each.

The "Electro" Junior Fixed Condenser

THE CONDENSER THAT IS COPIED BUT NEVER EQUALLED

The "Electro Junior" Condenser is the outcome of long experimenting and is the SMALLEST AND NEATEST wireless condenser ever placed before the public. It has the largest sale in the U. S. of a condenser of this class. It is entirely made of hard rubber composition and has hard rubber binding posts. Size over all 2⅝x1½ in., weight 3 ounces. This condenser is used mostly to shunt across the telephone receivers and is invaluable for any of the mineral detectors. We guarantee that the signals will come in fully 25 per cent. stronger with the addition of this condenser. A novel idea in connection with this instrument is that the diagram of connections is pressed right in the condenser top as seen in illustration. (Explanation of diagram (see ill.): T.T. telephone receiver (or receivers); C. "Junior" fixed condenser, arrows go to detector and battery, as the case may be.) Of course, the condenser can be used in other parts of the circuit and, two or three of these in any wireless station will prove a great addition, not alone on account of the neat appearance of the instrument but also on account of the increased receiving range of the station. The condenser itself is sealed in the case and there are no parts to get loose and form bad connections.

NO. FK10010

One use for which this condenser is particularly adapted is in the grid circuit of a vacuum type relay such as the Audion, etc. It is here that its small capacity and special construction make it particularly valuable. Here also does its superior dielectric prove valuable for the ordinary surge that occurs will never break down its insulation.

A comparatively recent development of the regenerative circuit such as the Armstrong makes use of small fixed capacities. For this purpose the No. FK10010 Junior Fixed Condenser is hard to beat. Its capacity is right and fixed and above all every condenser is exactly like every other one of its kind making them perfectly balanced and interchangeable.

Of course a very common use for the No. FK10010 Junior Fixed Condenser is as a blocking condenser where a very small capacity conveniently shaped and convenient for connections is required. It is then simply connected in series with the crystal detector.

Another use for this condenser that its low price and size make particularly useful is in conjunction with the test buzzer where it can be used to produce a better wave form.

This condenser will positively last a lifetime and cannot be punctured unless you connect it across the spark coil. CAPACITY is .0165 M. F.

No. FK10010 "Electro" Junior fixed Condenser, as described.... **$0.60**
Shipping weight 4 oz.

Gentlemen:—   West Hoboken, N. J.

Some time ago I purchased one of your "Interstate" receiving outfits, and I wish to say that this outfit HAS FAR EXCEEDED MY EXPECTATIONS.

A friend of mine purchased a tuning coil (a $2.00 coil) from another firm in New York City. His coil is four inches thick and a little longer than the one on my outfit and yet I can tune the Navy Yard, the Herald, and several other stations, BETTER ON MINE THAN ON HIS.

The "Interstate" is the BEST OUTFIT FOR THE MONEY that could be purchased.   Yours truly,   EDW. J. COTTERELL.

The "Electro" Rotary Variable Condensers

43 PLATE SIZE

$4⁰⁰

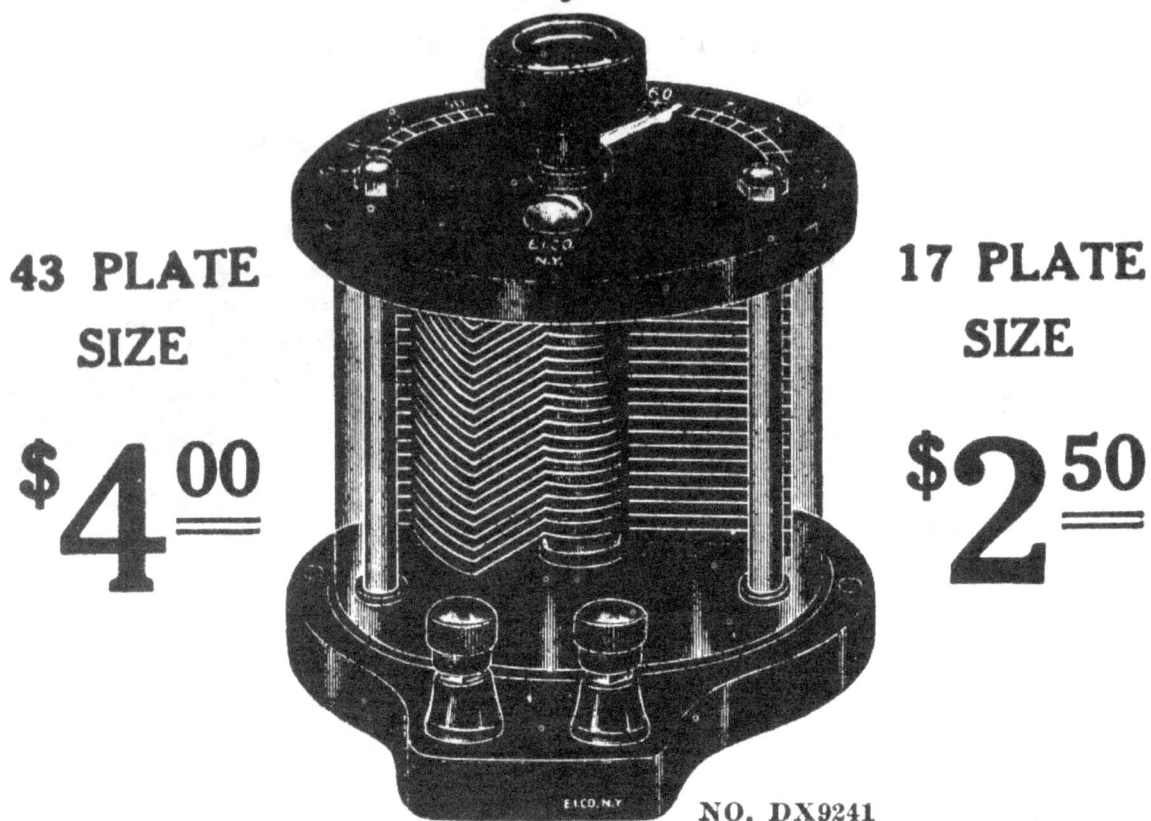

17 PLATE SIZE

$2⁵⁰

NO. DX9241

The best rotary variable condenser made.

While the rotary variable condenser of the slide plate **variety is not a** novelty to the wireless amateur it was never manufactured by us until all the faults were eliminated and a few virtues added. The rotary variable condensers we present here, have exclusive features which make them more valuable than others, yet, as usual, our price is lower. Consider these three features alone and you will be convinced: **FIRST—THESE CONDENSERS ARE THE ONLY ONES MADE WITH A TRANSPARENT CASE IN WHICH OIL CAN BE USED WITHOUT IT LEAKING.** In this way the condenser capacity can be increased **FIVE TIMES** and at the same time the condenser can be used on higher potential currents than air insulated condensers. Next, this condenser is the only one now on the market provided with screw holes so it can be screwed down to a table or instrument board. **THIRD—THIS CONDENSER IS THE ONLY ONE NOW ON THE MARKET WITH CONNECTIONS AT THE BOTTOM** as shown in the illustration. This form of construction costs us a whole lot more but it makes a better instrument and cleaner wiring for you. No longer is it necessary to run unsightly wires up to the top of your condenser, for our connections are only ½ in. above the table level. Cover is made of highly polished hard rubber composition with **a large scale that is easily read.** The handle is knurled and a very convenient size. The pointer is very rigid and clear and the handle has an exclusive feature permitting of it swinging all around in a complete arc or stopping at the maximum and minimum capacity. Plates are of a special metal alloy, properly spaced with separators milled to .0005 of an inch, and so supported that they can never slip or short circuit. The base is of one piece of hard rubber composition, with a beautiful finish that will stay on. The case is a special hard, clear flint glass cylinder, and by an exceedingly simple arrangement between same and the base, oil may be kept in it without the slightest possibility of leaking. Binding posts are our celebrated hard rubber which look fine, make perfect contact, yet cannot be short circuited.● If you want the finest rotary variable condenser ever built get the "Electro."

No. BEK9240 has 17 plates and a capacity of .0004 microfarads.
No. DX9241 has 43 plates and a capacity of .001 microfarads.

No. BEK9240 "Electro" Rotary Variable Condenser, 17 Plates, size 4⅛x2⅞ in.................................... **$2.50**

Shipping weight 2 lbs.

No. DX9241 "Electro" Rotary Variable Condenser, 43 Plates, size 4⅛x2⅞ in.................................... **$4.00**

Shipping weight 3 lbs.

The "Electro" Fixed Variable Condenser

THE BIGGEST VARIABLE CONDENSER VALUE IN THE U. S.

This is one of the greatest innovations ever originated by us. This condenser supersedes our No. ABE10000 wood case type and has the following striking improvements. The case contains two fixed condensers of different capacities. If the switch lever is on point 1 the two condensers are in series; this is the lowest capacity available. If lever is moved to point 2, the smallest condenser is in circuit. On point 3 the large condenser is placed in circuit. Thus it will be seen that three distinct capacities are provided for in this condenser. It is a proven fact that different stations are heard with varying degrees of intensity, all depending on the capacity of the ground (blocking) condenser. Very few persons realize that they cannot hear certain stations for the sole reason that their ground condenser is either too high or too low in capacity. For that reason this new style condenser was evolved by us and it has found the instant approval of thousands of wireless enthusiasts.

It is of especial value when used on regenerative wave circuits where a variable Condenser is desirable, yet fixed capacities are especially needed. It also fits in perfectly on small receiving outfits as either a primary or secondary loose coupler condenser. Its convenient shape and rotary system of operation is what will make its first appeal to you. Then will come its working qualities that will convince you that our claims are not exaggerated.

NO. ABE10000

This instrument is built on strictly scientific principles after the latest researches in condenser building. A special grade of dielectric is used and the capacities of the condensers are correct and balanced to meet all regular wireless requirements. The switching arrangement is unique and we absolutely guarantee that neither the switch lever nor the switch blade will come loose even through excessive use of the instrument. Contacts are of the self cleaning low resistance type.

The case is of solid hard rubber composition, as is also the thumb screw, which latter is polished; there are two stops to check the lever. Our illustration does the instrument but scant justice. It must be seen and worked to fully appreciate it.

Sizes are 4 in. x 1½ in. Shipping weight 1 lb.
No. ABE10000 "Electro" Fixed Variable Condenser............. **$1.25**

Dear Sirs:—   Vanceburg, Ky.

I received the motor all right and I like it very much, also "The Electrical Experimenter," which is a fine paper.   EDGAR PURDOM.

The Gernsback Rotary Variable Condenser

Patented July 23, 1912

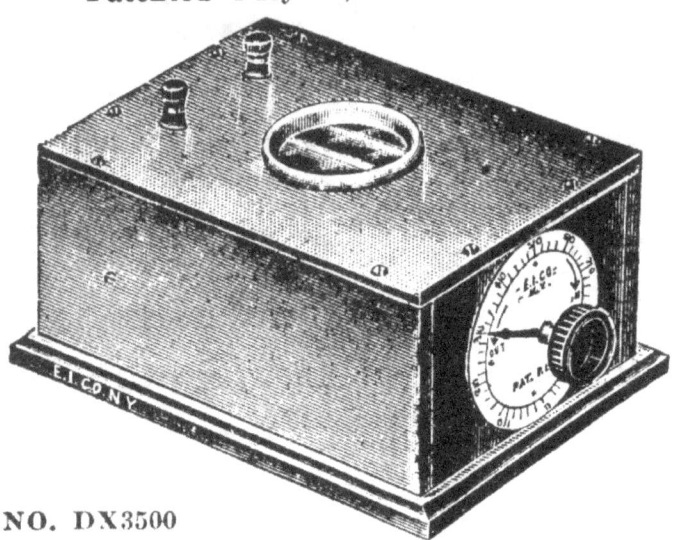

NO. DX3500

One of the most important instruments in the wireless receiving set is the rotary variable condenser, yet, it is strange to note that little attention and experimenting has been devoted to making it more perfect than rotary types using intersecting plates. Slide plate condensers have faults that are hard to eliminate, such as, short circuiting, jamming, low capacity and impossibility of getting zero capacity, inability to use in a horizontal position.

We therefore devoted our attention to the solution of the rotary condenser problem, and have evolved a new type, using an entirely new mechanical principle of operation eliminating ALL the defects of the rotary plate type. As will be seen by the illustration of the rotary condenser, the working parts are entirely encased in a neat oak box, finished in the natural grain and highly polished. A large rubber handle that is in the handiest position imaginable is used for manipulating the condenser, while a pointer indicates the proportion of the total capacity in circuit, on a neat scale. Two of our hard rubber binding posts are used on top of the box to make the connections to the receiving set. The apparatus being entirely contained in the oak case, makes the condenser dust-proof.

As for the working parts, it may be stated that these are far simpler than condensers using intersecting plates. The parts work in perfect unison, with no friction or opportunity for wear. It is impossible for the condenser to get out of order in consequence.

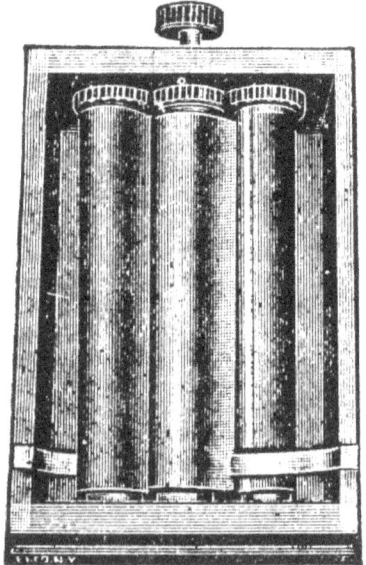

·INTERIOR VIEW

Among the many features embodied in our condenser, we may state the most important as follows: The working parts do not intersect, and nothing can bind or drop its adjustment. **The conducting surfaces being separated by one-thousandth part of an inch of dielectric material, give the condenser an extremely high capacity,** which is guaranteed by us to be at least **ten times that of any other variable condenser of equal dimensions.** The insulation is perfect, and breakdowns are impossible. However, it must be understood that this is a receiving condenser and not to be used on a transmitting outfit or other source of high tension current. It will not break down under static currents from the air which flow through the receiving apparatus. No plates with sharp edges being used, leakage is impossible. Short-circuiting of the condenser is impossible under any circumstances. The parts being simple and working with little friction or wear, give the apparatus a long life, far in excess of other condensers.

It is fool-proof, nothing being in sight or exposed so that it may be tampered with. **This variable condenser will work in any position.** We have made the apparatus with the handle on the side, so that **the arm of the operator may be rested** while adjusting the condenser, giving greater accuracy, and less effort in the manipulation. No tightening screw or nut is needed to hold the adjustment at one place. It is self regulating and holds any adjustment after the handle has been left at the point desired. No amount of shaking or vibration can affect the adjustment.

Finally, it is the best condenser at any price on the market, and you cannot afford to use the others if you wish the maximum efficiency.

As usual we lead—others follow—and copy.

Size, 9½x6x4⅞ inches. CAPACITY .01 M. F. $4.00

No. DX3500 Gernsback Rotary Variable Condenser (Patented).. $4.00

Shipping weight 7 lbs.

The "Electro" Professional Wave Meter

FOR WAVE LENGTHS FROM 180 TO 1,800 METERS

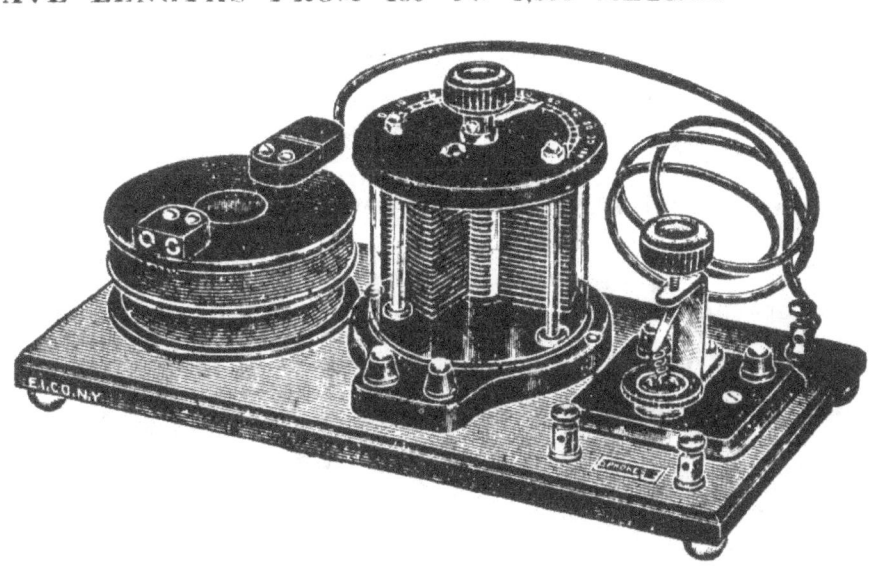

The up-to-date wireless amateur to-day wants more than a sending or receiving outfit or both. He wants to know what he is sending and what he is receiving. Realizing therefore the need of the radio enthusiast for an accurate measuring instrument we devoted a great deal of time and money to perfecting one that would produce the maximum of results with the simplest of instruments and with a maximum of accuracy even when in the hands of a mere novice.

Our Professional Wave Meter enables you to easily find out what wave length you are emitting and therefore to tune your station to comply with the law which requires an amateur station to use a wave length of 200 meters or less. The law goes further and says you must emit a wave with a decrement of 1/10 or less. Our wave meter enables you to so tune your station so it will emit a wave form acceptable to the government.

In other words, our Professional Wave Meter enables you to read wave lengths of either receiving or sending stations, also to obtain capacities, inductances and decrements, and then when you are through using your wave meter as such, JUST ADD A LOOSE COUPLER, TAKE OFF YOUR INDUCTANCE COIL AND YOU HAVE A FIRST CLASS RECEIVING OUTFIT.

WHAT IT CONSISTS OF:

Our Professional Wave Meter consists of two standard and accurately wound inductance coils on a seasoned and polished mahogany form having two neat separable connectors conveniently mounted for connections to either of the coils. This form is also called the exploring coil. When not in use it sets on a handy peg on the polished mahogany base. For connections we supply a 5 ft. silk cord. The detector is our standard No. AEK9701 RADIOCITE DETECTOR whose sensitivity is so well known that it requires no further mention. The condenser is our accurate and

When ordering one of our Wave-Meters, let us send you free with our compliments, lesson No. 14 "Operation of the Instruments" or lesson No. 19 "The Mathematics of Wireless Telegraphy" of our famous "WIRELESS COURSE" containing everything about Wave-Meters.

Just attach coupons No. 14 or No. 19 to your order. For further information see colored section of this catalog.

The "Electro" Professional Wave Meter
(Continued)

never varying No. DX9241 that is as near perfect as a condenser can be made. All are mounted on a beautiful hand rubbed piano finish mahogany base that will be an ornament to any station. The entire instrument rests on soft rubber feet for extra insulation. The directions that we supply are as complete as it is possible to make them and yet are so simple that they require no expert or trained user to get perfect results with the instrument. For readings we supply an accurate plotted curve that is of course absolutely essential. Altogether every part is of very high grade and assembled by expert mechanics so it will last and always be accurate and reliable.

OPERATION:

Do you want to find out what your emitted wave is? Simply bring the standard inductance near your sending helix or oscillation transformer. Move your condenser needle to the position where signals come in loudest in the receiver. Note the reading on your condenser and look for that reading on your curve **which immediately tells you the wave length in meters**, without lengthy mathematics. To read the wave length of an incoming wave bring the exploring coil close to your tuning coil or loose coupler and follow the same procedure.

Accuracy is guaranteed within 3 per cent.; sufficient for all commercial needs and surely for all amateur purposes.

Remember we were **not** the first to produce a wave meter, but we are the last. We have profited by the mistakes and experience of others and can therefore assure you of the best at the lowest price.

For those who desire to make their own wave meters we can supply the standard inductance separately, **but this is only of use when employed in conjunction with our No. DX9241 Condenser.** We disclaim all accuracy with other condensers. We urge, however, the purchase of the entire instrument complete. For receivers we advise the use of any of our better grade wireless receivers such as the Government or Transatlantic types, but any good wireless receiver will do excellent work. Receivers are not supplied with this wave meter.

Is your station up-to-date? If not bring it up-to-date by getting our Professional Wave Meter at once.

No. HX4488 "Electro" Professional Wave Meter, complete....... **$8.00**
    Size 7x14x6 in. Shipping weight 10 lbs.
No. BEK4489 "Electro" Professional Wave Meter Inductance only, with 2 windings on one form, (see illustration) with plug connectors and 5 ft. cord and 2 cord tips. Price....... **$2.50**
    Size 4x1⅝ in. Shipping weight 1 lb.

The "Electro" Ground Clamp

NO. AE10003

The most ingenious clamp ever invented. Invaluable to every wireless experimenter.

"A wireless outfit is not better than its weakest part"— which is usually a poor ground. Fifty per cent. of all wireless troubles are due to a poor ground. Our new Ground Clamp is, of course, not used only for wireless work, but for telephone, bells, telegraph and lighting work; in fact, everywhere where a good ground on which YOU CAN DEPEND is desired.

It fits any gas or water pipe from ½ in. to 2 in. diameter. A tinned lug is provided to attach wires. Clamp is installed in less than two minutes. Tools needed to install: A screwdriver —that's all! No wires to be wrapped around pipes, which method **always** gives trouble. The contact-band of our clamp (9/16 in. wide) is of pure copper with the lug tin plated.

No. AE10003 "Electro" Ground Clamp, as described, each...... **$0.15**
    Shipping weight 4 oz.

The "Electro" Radiotone
HIGH FREQUENCY SILENT TEST BUZZER

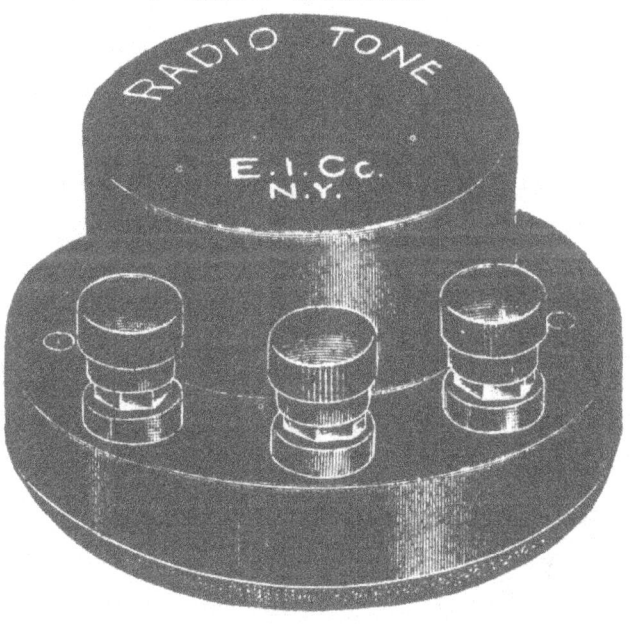

NO. IK1800

The "Electro" RADIOTONE was designed and constructed by us first in 1912, but it did not quite come up to our expectations and for that reason was not listed by us until very recently. Four years of experimenting finally brought this wonderful instrument to such a high state of perfection that we now have no hesitancy in proclaiming it the most perfect as well as the most efficient instrument of its kind on the market to-day, **irrespective of price.**

The RADIOTONE is NOT a mere test buzzer, it is infinitely more. Mr. H. Gernsback who designed this instrument labored incessantly to produce an instrument which would imitate the sound of a high power Wireless station as heard in a set of phones. This actually has been achieved in the RADIOTONE.

This instrument gives a wonderful high pitched MUSICAL NOTE in the receivers, impossible to obtain with the ordinary test buzzer. The RADIO-TONE is built along entirely new lines; it is NOT an ordinary buzzer, reconstructed in some manner. The RADIOTONE has a single fine steel reed vibrating at a remarkably high speed, adjusted to its most efficient frequency at the factory. Hard silver contacts are used to make the instrument last practically forever. There is nothing to get out of order —for there are no set screws, no adjusting screws, which in themselves proclaim an instrument as unperfected.

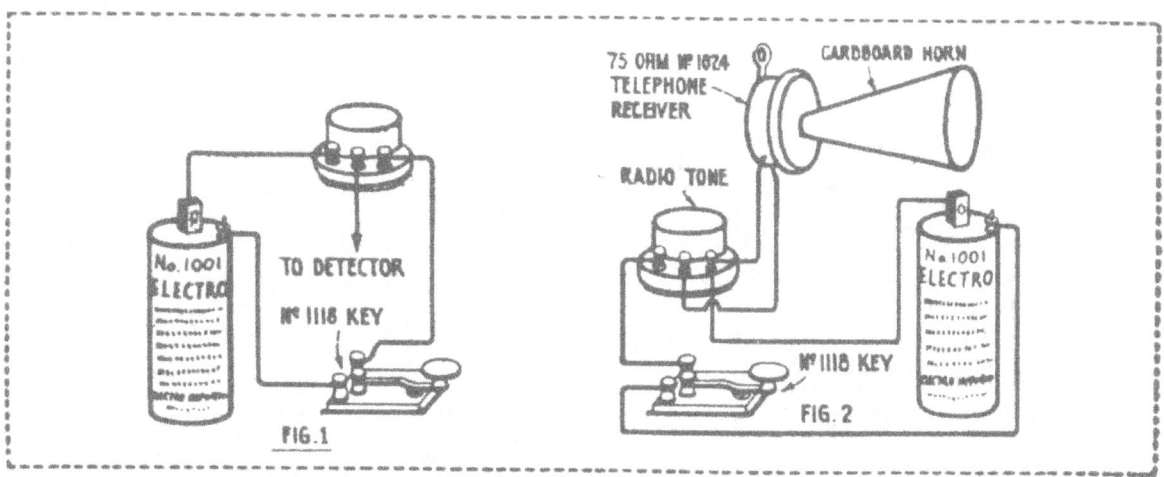

Yes, the RADIOTONE is **SILENT**. In fact, it is so silent that you must place your ear almost on top of it to hear its beautiful musical note. If you have to adjust your detector you appreciate just what this means. Nearly all test buzzers on the market to-day, scream so loud that you hear them 15 feet and more away. How can you adjust a detector, when you hear TWO SOUNDS, one outside of the phones, the other inside of the phones? Nothing like this with the RADIOTONE. You hear the sound where it belongs—in the phones.

How do we do it? First the steel reed is so constructed that it can not possibly create a loud sound in the air surrounding it. Then by acoustically insulating the entire electrical unit, and by providing a heavy felt base for the instrument, all outside sound is done away with.

The "Electro" Radiotone (Continued)

The casing is made of hard rubber composition and there are three of our well known hard rubber composition binding posts. The ones for the battery connection are black, the one for the detector RED—a simple refinement, but important to show E. I. Co. attention to details.

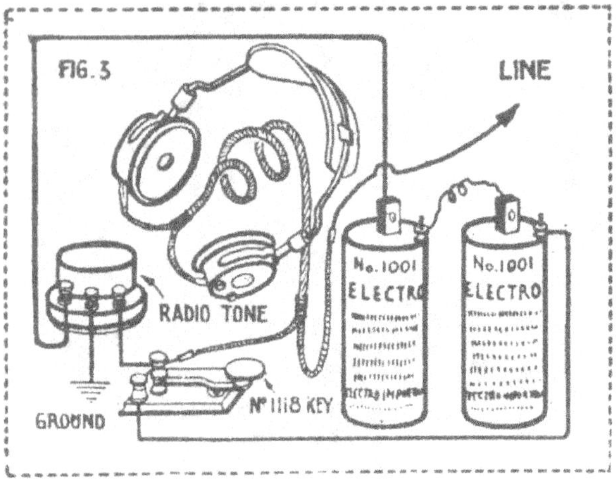

Fig. 3

Then too a big feature—COM· PACTNESS. The RADIOTONE is small and takes up but very little room. Just the same, we wager you will give it a prominent location on your instrument table, because it really is an exceptionally beautiful instrument, one you will be proud to show to your most critical friends.

As already mentioned the RA-DIOTONE is equipped with an exceptionally heavy green felt sub-base. This gives the instrument a very distinctive appearance.

The RADIOTONE works best on a single dry cell. Two cells may be used but we do not recommend this. The RADIOTONE can be operated continuously if desired, it will POSITIVELY NOT STICK as do so many test buzzers on the market to-day, and which must often be adjusted.

RADIOTONE LEARNER'S OUTFIT

No instrument lends itself more readily towards learning the telegraph codes than the RADIOTONE. An ordinary telegraph sounder outfit is worse than useless to learn the wireless codes because every time you depress the key and hold it down no sound is heard but the first click. It does not resemble in the least the sounds heard in a set of phones when receiving a Radio message.

The RADIOTONE, however, lends itself admirably to this purpose. It gives an **exact reproduction** of a Radio message and you can readily learn the codes in less than thirty days with only a little persistent practicing.

Fig. 2 shows what a real learner's Radio Code Outfit consists of: You require first the RADIOTONE; second a dry cell; third our No. EK1024 Receiver (75 ohms); fourth our No. CE1118 Telegraph Key; fifth our No. FK10010 Condenser. A few extra receivers may be connected as shown by dotted lines, in case several of your friends are learning the code with you. Don't forget: in all cases the condenser MUST be used.

If you wish comfort, order one of our No. AX8077 headbands and an extra receiver, to keep the receivers to your ears.

INTERCOMMUNICATING RADIOTONE OUTFIT

Fig. 3 shows another suggestion for a modern telegraph line, to practice telegraphy between two chums' houses.

As will be noted but one metallic line wire is required. The return circuit may be the ground as indicated. Each station consists of one RADIOTONE, one or more dry cells (according to distance); one of our No. CE1118 telegraph keys; one No. FK10010 Junior Fixed Condenser; two No. EK1024—75 ohm receivers (of course a single receiver may be used); one No. DE8075 5-foot receiver cord, and one No. AX8077 headband.

It will be noted that no current flows when the keys are at rest, and no switches are required. A call bell is not required as the phones will sing so loud that the tone may be heard ten feet away.

As a rule young experimenters get little pleasure from the old-fashioned sounder telegraph sets, because they are too noisy and parents usually object to the incessant **rat-tat-tat**.

No such objection to the RADIOTONE outfit. It is silent for all, except for yourself and your chum.

Size of instrument over all 2¾x1⅝ in.

No. IK1800   The "Electro" RADIOTONE, as described......... $0.90

Shipping weight 1 lb.

"Electro" Wireless Telephone Receivers

VERY few people realize that the wireless telephone receiver, without exception, is the most important part of a wireless receiving set. A sensitive detector is practically rendered useless if used in connection with a poor set of phones, while on the other hand a detector of but poor sensitivity, will sometimes achieve wonders with a high grade set of phones.

Many amateurs and most professionals know this fact only too well, and no matter how poor the rest of their receiving outfit, their phones will usually be found to be a good set.

Buying phones is much like buying clothes—you can buy both at most any price. In each case you get exactly what you pay for. It's the experience first and then the material which counts. Then, also the reputation of the firm. We could go on a long tirade telling you how many governments, and universities use our receivers **EXCLUSIVELY**, we could dazzle you with phenomenal figures of our output in telephone receivers, we could publish bushels of testimonials from satisfied customers, but we prefer to tell you how our phones are made. It probably will convince you better than anything else.

MATERIALS

The vital items in any receiver are the following: The permanent magnets, the wire used, the pole pieces, the diaphragms.

The permanent magnets in all our wireless receivers are made of the highest grade tungsten steel, manufactured for us expressly in Sweden and imported by us. After the magnets are blanked out to shape they are hardened glass hard. There is a big waste in this process as many pieces warp so badly that they cannot be used. Each blank is then tested for hardness by an inspector and no blank that can be scratched by a file is acceptable. Such blanks are scrapped. Consider that this steel costs us from 35 to 40 cents per lb. before blanking, when domestic steel can be had for from 16 to 20 cents. Then consider that but 70 per cent. of the blanks pass inspection, and that a pound of the steel does not furnish many blanks either.

The pole pieces, which perform a very important function, must be of the softest possible iron. Nothing but the best grade of imported Swedish iron will do, and not all grades of it either.

The wire used in all our receivers now, is imported enamel wire. For our high grade receivers we use No. 43 B. & S.—a wire so fine that the eye hardly perceives it. Only a specially trained operator can wind it on account of its great fragility.

There is at present—we are sorry to say—no suitable No. 43 enamel wire made in the United States. None seem to have the high insulating values of the European wire and the domestic wire of this size, furthermore, tears so easily that it cannot be used in a high speed winding machine.

There have been some controversies as to the use of enamel wire for wireless receivers in the past. To this let us say that up to 1909 we used single silk covered wire exclusively. Exhaustive comparative tests have shown that enamel wire is infinitely more efficient, and since 1909 we have used it almost exclusively. Since 1912 we have used nothing but enamel wire windings. The highest American, as well as European authorities have long come to the same conclusion; all European radio re-

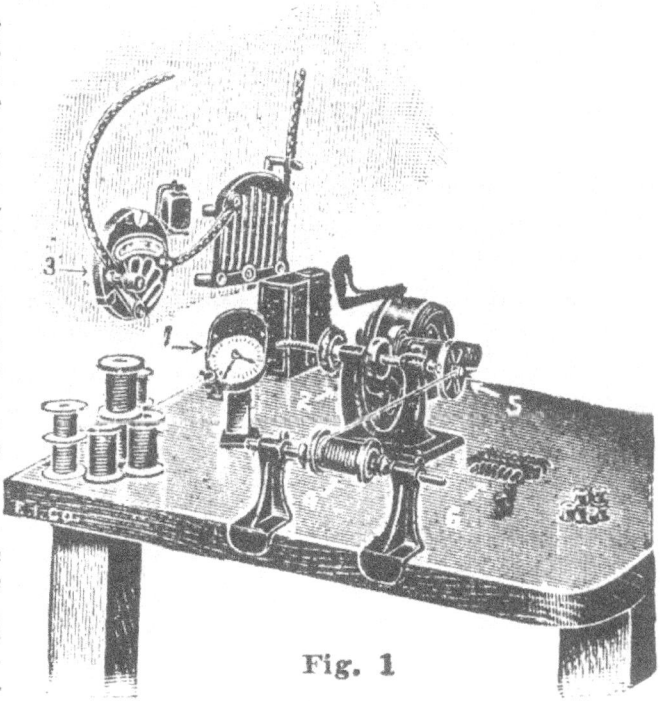

Fig. 1

ceivers - are now wound with enamel wire. We cannot here go into a lengthy technical discussion as to the great merits of enamel wire; suffice it to cite

the following: No one will deny that the more (ampere) turns one can wind on a given core, the better the electro-magnetic results, always bearing in mind, of course, that the nearer the windings to the core the more efficient the electro-magnet will be. Thus, in 1909, each one of our No. FGE1305 receiver spools were wound with 5,700 turns of No. 44 B. & S. silk wire. We now use 6,600 turns of No. 43 B. & S. wire—a heavier wire (carrying more current) and winding almost 1,000 more turns wire in the same space, without increasing the distance between the pole piece and the last layer. This example makes it clear why our receivers are now so infinitely more sensitive than they were in 1909. It also explains why they are more sensitive than most other makes.

"Resistance" is a very empty quotation in connection with a radio receiver. Thus, if we could use German silver wire in our receivers we would save several thousand dollars a year. Of course, it is absolutely impossible to use this wire (although some unscrupulous makers are still doing it), for the reason that resistance wire "chokes" almost all signals. We are not even satisfied to use good copper wire, hence our specifications for enamel wire invariably call for the highest grade electrolytic copper. This costs quite a little more, but we get from 1 per cent. to 3 per cent. better conductivity—an appreciable amount when we wind over 6,000 turns of wire on a bobbin.

The diaphragm is one of the least understood parts in telephone engineering to-day. The highest authorities have written books and pamphlets about it, there have been endless discussions, and it is quite safe to say that some millions of experiments have been made, in order to find out just how a diaphragm works. We only know this: We know mighty little as yet. In Mr. Gernsback's laboratory can be found several hundred diaphragms made of strange materials, as well as in strange shapes. There are diaphragms of pure nickel, some of the new magnetic alloys, some made of the latest silicon-transformer iron, some with curious holes and curious laminations, some made of iron wire, some made of mica with iron center, etc., etc. Our experience leads us to doubt if there is a concern in the country who lays greater stress on a diaphragm than we do. For our better receivers each diaphragm is hand selected and tested to achieve greatest uniformity. It also undergoes several other special tests, which, being trade secrets, cannot be mentioned here.

WINDING

The most important as well as interesting operation is the winding of our telephone bobbins. A specially constructed machine does this work, having been constructed in our own shops for our requirements. Fig. 1 shows it. 1 is the automatic revolution counter, which counts the turns of wire on each bobbin. Each style receiver bobbin is wound to a certain number of turns. Thus each bobbin of our "Government" receivers is wound to 6,600 turns, while each bobbin of our "Transatlantic" receivers is wound to 5,500 turns and so on. When the hand of the dial arrives at the, say, 6,600th turn, a bell rings and the operator stops. As our wire is remarkably uniform, the resistance of each bobbin will not vary for more than 1 to 2 ohms. The winding machine proper, 2, is operated by an electric motor. The control of the machine is foot operated. The speed regulator is shown at 3. The fine wire supply-spool is shown at 4, while the bobbin, 5, in process of winding, is held in an ingenious chuck. Completed bobbins are shown at 6.

Only a thoroughly experienced female operator can wind the No. 43 wire at a high speed without having several breaks in each bobbin. It is a very hard task at best, very trying as well as tiring. It is also necessary to operate the winder at a high rate of speed as else the wire does not "pack" tight enough. Thus an inexperienced operator can only get about 75 per cent. of the required wire on its bobbin. Naturally the machine must run very steady and smooth to obtain the correct results. After the bobbins are wound and equipped with lead wires, they are tested by means of an ohm meter. Those that test either too high or too low are discarded. The bobbins next go to the female assemblers who make up the complete receiver. Before they leave the assemblers' hands, however, the receiver undergoes various tests and several inspections.

CALIBRATING

The receivers are now placed in trays, each tray containing some 60 receivers. They are then taken to the calibrating machine, where the pole pieces, as well as the receiver shell edge, are machined down within 1/5000 of an inch accuracy. For it is of the utmost importance that the pole pieces are of quite equal height, and that the edge of the casing be a certain few thousandths of an inch higher than the pole pieces. Also the rim of the casing must be absolutely smooth and even as well as absolutely parallel with the pole pieces. This calibrating is a very tedious operation and can only be performed by skilled mechanics having long experience. This calibration process takes from 8 to 10 minutes for each receiver. After the receivers have been calibrated they are inspected by means of a micrometric appliance to make sure that the pole pieces come to 1/5000 of an inch parallel with the rim. If the variation is too great, the receiver must be recalibrated.

MAGNETIZING

After calibrating, the receiver is ready to receive its magnetic baptism. Fig. 2 shows the apparatus that does it. The extremely powerful electro-magnets, built in our own shops, connect without any resistance to the 220 volt power current. A large manipulating key, 2, serves to close and break the current, which circulates around the spools, 1. When energized, this electro-magnet **IS CAPABLE OF LIFTING OVER 500 LBS.** Only a very powerful electro-magnet produces satisfactory receivers.

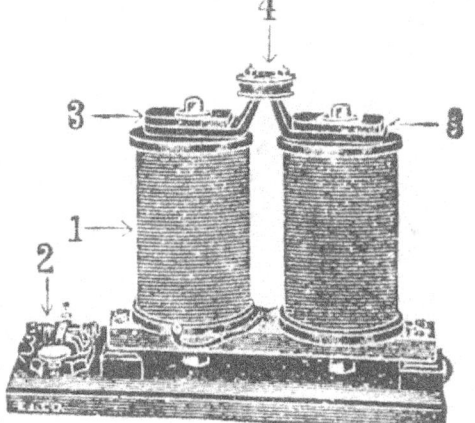

Fig 2

On top of the spools we see the adjustable pole pieces 3, 3. On top of these the telephone receiver, 4, is now placed by the magnetizing operator. The current is then turned on and broken a number of times, in a peculiar manner, while the receiver undergoes several mechanical treatments, which, being a secret process, cannot be mentioned here. After a few minutes, the receiver is completely magnetized, and is now tested once more, this time to ascertain if it is magnetically strong enough. The receiver, if passed by the inspection, now goes back to the assembling department. Here it is cleaned thoroughly and polished. Certain parts inside are then enameled and lacquered. After drying, the diaphragm, as well as the hard rubber cap, is put in position and the receiver is finished. The receivers are then assembled on the head band, the cords are put on and the head set is ready for the final tests. The first test is for resistance. If this has been found correct, the completed receivers are sent to the wireless room, where, by means of several ingenious instruments, the phones undergo three tests for sensitivity. Those not checking up with the "standard" are returned to the factory.

The final, supreme test is the wireless test in actually receiving signals. Phones not sounding clear and sharp are returned to the factory to be made over.

This completes the receivers. Each set is now tagged and signed by the tester and the phones are then ready for packing.

GUARANTEE

All our receivers are guaranteed for one year. If, during this time, they do not prove entirely satisfactory, for any reason whatever, we will exchange the set unhesitatingly for a new one, without question on our part, paying transportation charges both ways. If we did not think that our phones were the best in the country, we couldn't afford to make such a sweeping guarantee.

The Electro "Government" Phones

Highest Precision Phones Made in the United States
(Adopted by several Governments)

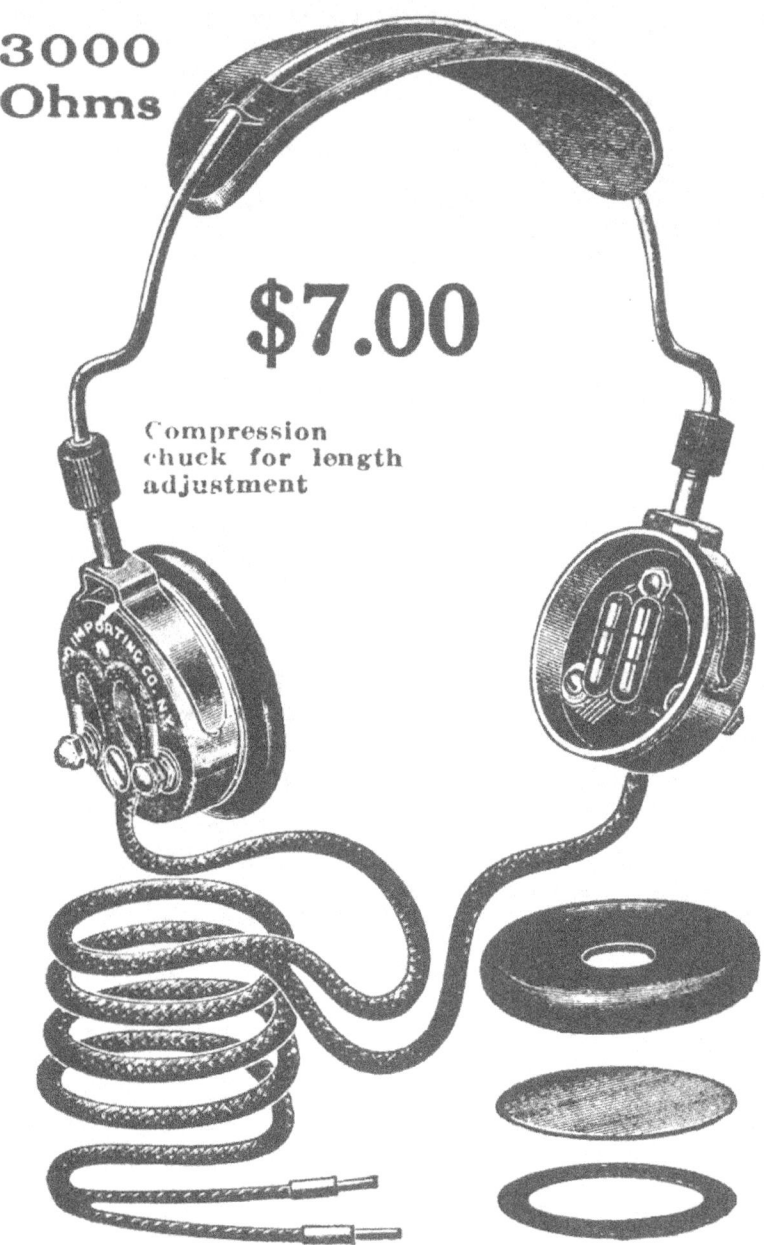

3000 Ohms

$7.00

Compression chuck for length adjustment

No. GX6666 (Patent Pending)

We have sold more wireless Telephone Receivers during the past twelve years than all the rest of our competitors combined, and while we thought that our No. GX1305 Phones could not be improved in any way, we have had right along a demand for a higher price receiver and after experimenting for some years we have developed a high grade professional type which eclipses anything shown heretofore.

We still maintain that our No. GX1305 Phones cannot be matched for the price anywhere and to-day the same as years ago, they hold their own against receivers selling from $13.00 all the way up to $20.00 a set.

The difference between our "Government" phones and the GX1305 type is merely in workmanship. The receiver shells are made of Aluminum, which makes them considerably lighter than other sets.

The magnets are wound with No. 43 B. & S. ENAMEL COPPER WIRE. The magnets are a great deal more powerful, being made of the best imported Swedish tungsten steel, which we guarantee. The Magnetic power of this receiver is the highest of any and this accounts for the remarkable sensitivity and long distance receiving power of these wonderful phones.

We lay particular stress on the magnets of this receiver and we guarantee that the magnets will not lose their strength for two years. We realize a **wireless telephone receiver is not better than the strength of its magnets**, hence we have extended all our energies towards producing something that can be relied upon, practically indefinitely.

While this headgear is the lightest on the market to-day we have not sacrificed its efficiency, as may be easily ascertained when testing out the sets. **Each receiver is wound to 1,500 ohms, giving 3,000 ohms per set.** Non-rusting diaphragms are supplied.

The Electro "Government" Phones
(Continued)

These phones as well as all our others are now equipped with our famous "**Gernsback Patent**" **Common-Sense Headbands.**

Our Mr. H. Gernsback had been experimenting for years before this extraordinary simple as well as efficient headband was finally developed.

Greatly annoyed by headbands that would not fit the head permanently, that would not hold the receivers tight to the ears, that caught your hair, that were heavy and hurt your head, he developed the present band that has none of these faults.

It does all this and then some:

1° Will not catch and tear your hair; as ALL doubleband headbands do.

2° Utmost comfort assured—molded soft rubber pad does it.

3° Fits any head instantly. Can be shortened or lengthened simply by unloosening chucks.

4° Lightest band on the market—weight 5 oz.

5° Fits the receivers to your ears perfectly and keeps them there excluding all outside noises.

6° Band on head, is almost invisible, consequently not unsightly as are ALL others.

7° No metal touches your head—no shocks, no leakage.

8° Has less parts than ANY other band, consequently gives less trouble.

9° The powerful Hard Brass spring wire keeps the 'phones pressed to your ears, with an even pressure ALWAYS. You can't possibly shake the 'phones from your head.

10° Beautiful hand-buffed nickel finish. Sanitary soft rubber pad.

Until you have worn a "**Gernsback Patent**" **Common-Sense Headband** you don't know what 'phone comfort is.

Other points of superiority: Light rubber cap, highest insulation. The phones will fit the ear snugly to exclude all external noise. We use a five-foot pure silk covered cord with two tips. As a remainder always be sure with whom you deal. Anyone can make extravagant, untruthful claims. Anyone can imitate us, **for a certain time.** Before you buy from any concern find out how long they have been in business. If you don't you will buy their EXPERIMENTS, not a product that has been on the market for over a decade as, for instance, ours.

Finally, don't forget that the firm who "knocks" us most is so busy thinking about us that it hasn't got much time to accomplish anything worth while itself. Now that you grasp the situation read our

GUARANTEE

Buy a pair of these phones and if it is not as sensitive or not as satisfactory as any set you have ever seen, or if it does not compare in efficiency with any other make, return it to us and we will give credit for same at once.

No. GX6666 Electro "Government" Phones, as described **$7.00**
Shipping weight 2 lbs.

No. BHE6667 Single Receiver (no band or cord), 1,500 Ohms **$2.85**
Shipping weight 1 lb.

WE WIND THESE RECEIVERS TO ANY RESISTANCE. ASK FOR PRICES

When ordering any of our Phones, don't forget that we will gladly send you FREE with our compliments lesson No. 2 "**The Principles of Magnetism**" or lesson No. 9 "**Receiving Apparata**" or lesson No. 18 "**The Wireless Telephone**" of our famous "WIRELESS COURSE."

Everything worth knowing about Receivers is explained in these lessons.

Just attach your free coupons to your order. For further information consult colored section of this catalog.

The "Electro" Transatlantic Phones

The Phones Which Saved the "Republic"
(or rather its 500 human beings)

Operator Binns, the famous C. Q. D. man, used these Phones.

HIGH RESISTANCE PRECISION HEAD RECEIVERS FOR WIRE-LESS TELEGRAPHY AND TELEPHONY, Transatlantic Type.

Adopted by the U. S. Navy, United Wireless Co., etc.

These receivers embody the finest workmanship, and in connection with our various Detectors and other instruments are so marvelously sensitive that they will talk loud and distinct where others will not respond.

We make the broad and sweeping statement that our receivers are absolutely the most sensitive in the world now—without any exception and regardless of price. We have hundreds of testimonials from enthusiastic owners of our headphones.

The weight is 20 per cent. less than other similar receivers; operators do not tire with these even if worn hours at a stretch.

These phones as well as all our others are now equipped with our famous "Gernsback Patent" Common-Sense Headbands.

Our Mr. H. Gernsback had been experimenting for years before this extraordinary simple as well as efficient headband was finally developed.

Greatly annoyed by headbands that would not fit the head permanently, that would not hold the receivers tight to the ears, that caught your hair, that were heavy and hurt your head, he developed the present band that has none of these faults.

It does all this and then some:

1° Will not catch and tear your hair; as ALL doubleband headbands do.

2° Utmost comfort assured,—moulded soft rubber pad does it.

3° Fits any head instantly. Can be shortened or lengthened simply by unloosening chucks.

4° Lightest band on the market—weight 5 oz.

5° Fits the receivers to your ears perfectly, and keeps them there, excluding all outside noises.

6° Band when on head, is almost invisible, consequently not unsightly as are ALL others.

7° No metal touches your head—no shocks, no leakage.

8° Has less parts than ANY other band, consequently gives less trouble.

9° The powerful German silver spring wire keeps the 'phones pressed to your ears, with an even pressure ALWAYS. You can't possibly shake the 'phones from your head.

10° Beautiful hand-buffed nickel finish. Sanitary soft rubber pad.

Until you have worn a "Gernsback Patent" Common-Sense Headband you don't know what 'phone comfort is.

For illustration of how these phones look see illustration for Receivers No. GX6666.

Each receiver is wound to 1,000 OHMS with No. 41 B. & S. Enamel Electrolytic copper wire which explains the extraordinary sensitivity.

This fine wire costs six times as much as other wires, but we use it because we increase with its use the AMPERE TURNS and the receivers consequently become infinitely more sensitive. We guarantee each receiver to stand the following extraordinary test: Moisten or wet the metal receiver cord tips. When both are touched the receiver will respond! The voltage generated by the metal tips is less than 1/1000, the amperage less than 1/1,000,000 (one millionth). Bands and receivers finely nickel plated. **NON-RUSTING DIAPHRAGMS.** Silk conductor cords, 5 feet long.

No. FX1305 Head Receivers (2) with head band............... **$6.00**
Shipping weight 2 lbs.

No. 3CE1307 Receiver only with 3 foot cord............... **$2.35**
Shipping weight 1 lb.

No. AX1308 Double Head Band only (no cord)............... **$1.00**
Shipping weight 1 lb.

No. AX1024a 1,000 Ohm Single Pole Receiver............... **$1.00**
Shipping weight 1 lb.

WE WIND THESE RECEIVERS TO ANY RESISTANCE WANTED. WRITE FOR PRICES.

The "Electro" Amateur Wireless Phones

2,000   Ohms

$4.00

Compression
Chuck for length
adjustment

We herewith present our amateur type wireless phones which are superior to anything as yet. Our No. FX1305 phones, which are in use now by the United States Government, etc., are, of course, a higher grade but our amateur phones are in every respect built as carefully, the only difference being that the finish is not so elaborate. These phones are wound to 1,000 ohms each receiver and are wound with No. 40 enamel copper wire. These phones have double pole magnets, which are extremely powerful and made especially for wireless.

These phones as well as all our others are now equipped with our famous "Gernsback Patent" Common-Sense Headbands.

Our Mr. H. Gernsback had been experimenting for years before this extraordinary simple as well as efficient headband was finally developed.

Greatly annoyed by headbands that would not fit the head permanently, that would not hold the receivers tight to the ears, that caught your hair, that were heavy and hurt your head, he developed the present band that has none of these faults.

It does all this and then some:

1° Will not catch and tear your hair; as ALL doubleband headbands do.

2° Utmost comfort assured — moulded soft rubber pad does it.

3° Fits any head instantly. Can be shortened or lengthened simply by unloosening chucks.

4° Lightest band on the market—weight 5 oz.

5° Fits the receivers to your ears perfectly and keeps them there excluding all outside noises.

6° Band when on head, is almost invisible, consequently not unsightly as are ALL others.

7° No metal touches your head—no shocks. no leakage.

No. DX8070 (Patent Pending)

The "Electro" Amateur Wireless Phones
(Continued)

8° Has less parts than ANY other band, consequently gives less trouble.

9° The powerful Hard Brass spring wire keeps the 'phones pressed to your ears, with an even pressure ALWAYS. You can't possibly shake the 'phones from your head.

10° Beautiful hand-buffed nickel finish. Sanitary soft rubber pad.

Until you have worn a "Gernsback Patent" Common-Sense Headband you don't know what 'phone comfort is.

The receivers fit the head perfectly. The weight is 12 ounces. With this set we furnish a finely finished five foot bifurcated green cord with nickel-plated tips. A test will convince you that our phones are superior to any other make and if they are not exactly what we claim them to be we shall refund the money.

No. DX8070 Two Thousand Ohm Phones, as described **$4.00**
Shipping weight 2 lbs.

No. AEK8071 Receiver only (1,000 Ohms), as furnished with No. DX8070 (double pole) **$1.50**
Shipping weight 1 lb.

No. AX8077 "Gernsback Patent" Double Headband (fits our No. EK1024, AX1024a, AEK8071) **$1.00**
Shipping weight 1 lb.

No. DE8075 5 foot bifurcated green cord, each **$0.45**
Shipping weight 4 oz.

THE "ELECTRO" "JUNIOR"
Wireless Phones

2,000 Ohms

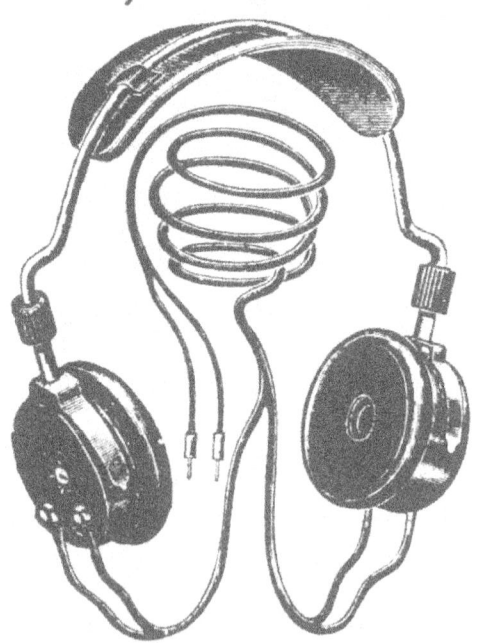

No. CX8090 (Patent Pending)

These phones are exactly the same as our No. DX8070 described above with the exception that instead of using No. AEK8071 double pole receivers, we furnish 2-1,000 ohm No. AX1024a single pole receivers. While these are single pole receivers, it should be borne in mind that in connection with silicon or galena detectors, such phones will almost prove as sensitive as the No. DX8070 kind. These phones are marvelously sensitive and will give a click when the two moistened tips are contacted with another piece of metal,—a test which very few high priced receivers will stand.

These phones as well as all our others are now equipped with our famous "Gernsback Patent" Common-Sense Headbands.

Our Mr. H. Gernsback has been experimenting for years before this extraordinary simple as well as efficient headband was finally developed.

Greatly annoyed by headbands that would not fit the head permanently, that would not hold the receivers tight to the ears, that caught your hair, that were heavy and hurt your head, he developed the present band that has none of these faults.

IT DOES ALL THIS AND THEN SOME:

1° Will not catch and tear your hair; as ALL doubleband headbands do.

2° Utmost comfort assured—molded soft rubber pad does it.

3° Fits any head instantly. Can be shortened or lengthened simply by unloosening chucks.

4° Lightest band on the market—weight 5 oz.

5° Fits the receivers to your ears perfectly and keeps them there excluding all outside noises.

6° Band when on head, is almost invisible, consequently not unsightly as are ALL others.

7° No metal touches your head—no shocks, no leakage.

8° Has less parts than ANY other band, consequently gives less trouble.

9° The powerful Hard Brass spring wire keeps the 'phones pressed to your ears, with an even pressure ALWAYS. You can't possibly shake the 'phones from your head.

10° Beautiful hand-buffed nickel finish. Sanitary soft rubber pad.

Until you have worn a **"Gernsback Patent"** Common-Sense Headband you don't know what 'phone comfort is.

The "Electro" Junior wireless phones consist of two receivers. "Gernsback Patent" swivel soft rubber pad headband and five foot bifurcated cords are furnished with this set.

No. CX8090 2,000 ohm Junior Wireless Phones, as described per set **$3.00**

Shipping weight 2 lbs.

The "Electro" Double Pole Receiver

The double pole receiver No. HE1030, which we illustrate here is of course more powerful than the No. EK1024 type. The No. HE1030 receiver is wound to 75 ohms and has two powerful magnets and double poles. This telephone receiver is suitable for all kinds of telephone work where a powerful double pole receiver is wanted and is also found of great use in wireless telegraphy where a low resistance receiver is desired.

Of course, it can be used in the same way as our No. EK1024 but in all cases it will give better results just as you have a right to expect for it is double pole and stronger.

One of the uses many of our customers have been putting it to is in the making of microphonic telephone and wireless amplifiers using it as a second step up from the transmitter used. In that case a high grade low resistance receiver is required and high grade receivers are the only kind we make.

NO. HE1030

It can also be used as a second receiver for regular telephones in that way providing one receiver for each ear. Try it once and you will always have it on your phone. Keeps out outside noises and lets you hear on both ears as nature intended. Simply connect it in parallel to your present receiver using one of our single receiver cords.

Sizes 2½x1¼ inches.

No. HE1030 75 Ohm Receiver, as described. Price................ **$0.85**

Shipping weight 1 lb.

Dear Sirs:—   Hightstown, N. Y.

I lately purchased a pair of your 2,000 ohm Transatlantic type phones, and they work great. The signals come in twice as loud as my others. I purchased the phones in your store, 69 West Broadway, N. Y.

LE ROY WEST.

The "Electro" 1000 Ohm Single Pole Receiver

$1.00

each

1,000

Ohms

THE BIGGEST WIRELESS RECEIVER VALUE IN THE COUNTRY

It isn't often that we can offer you so valuable a piece of apparatus at so little money. Here is a case in point.

Our No. AX1024A is a carefully designed wireless receiver of the single pole type and one which with certain types of detectors such as the silicon, galena, carborundum, etc., will give excellent results. This is not a cheap telephone receiver wound to a high resistance and then called a wireless receiver, but it was actually designed for the purpose we advertise.

To make you fully appreciate what this wonderful article is, let us tell how it is made and you judge for yourself.

The shell is of polished hard rubber composition, light, strong, and durable. The earpiece is of the same material and designed to be comfortable yet exclude external noises. Shell fits our regular headband. The magnet is a very fine special tungsten alloy magnet steel, very ingeniously shaped. It will retain its magnetism under all conditions, short of abuse.

The winding is a full tested 1000 OHMS IN NO. 40 B. & S. BLACK ENAMELED WIRE. It is wound on a specially soft Swedish iron core. Every receiver is tested for resistance and insulation. Diaphragms are of selected stock and hand sorted.

After reading this description do you doubt that we are proud of our No. AX1024A Receiver? You will never regret buying one or a pair. Their sensitivity far excels many double pole receivers now offered as "remarkably sensitive." Size 2½x1½ inches.

No. AX1024A   Receiver 1,000 ohms as described. Price.......... **$1.00**
Shipping weight 1 lb.

Gentlemen:—   Colma, Cal.

I received order No. 1,052,470 and find the phones just as described in your catalog. They fit the head perfect and are very sensitive. I find that the E. I. Co. is honest and that IT PAYS TO DEAL WITH YOU. You will find that I will patronize the E. I. Co. for all the goods I buy, wireless or raw materials.   Yours truly,   D. M. KUNTZEN.

A complete chapter on "TELEGRAPHS AND TELEPHONES" is contained in the "EXPERIMENTAL ELECTRICITY COURSE" in 20 lessons which is given FREE with one year's subscription to the **"Electrical Experimenter Magazine."** See announcement on back cover of catalog.

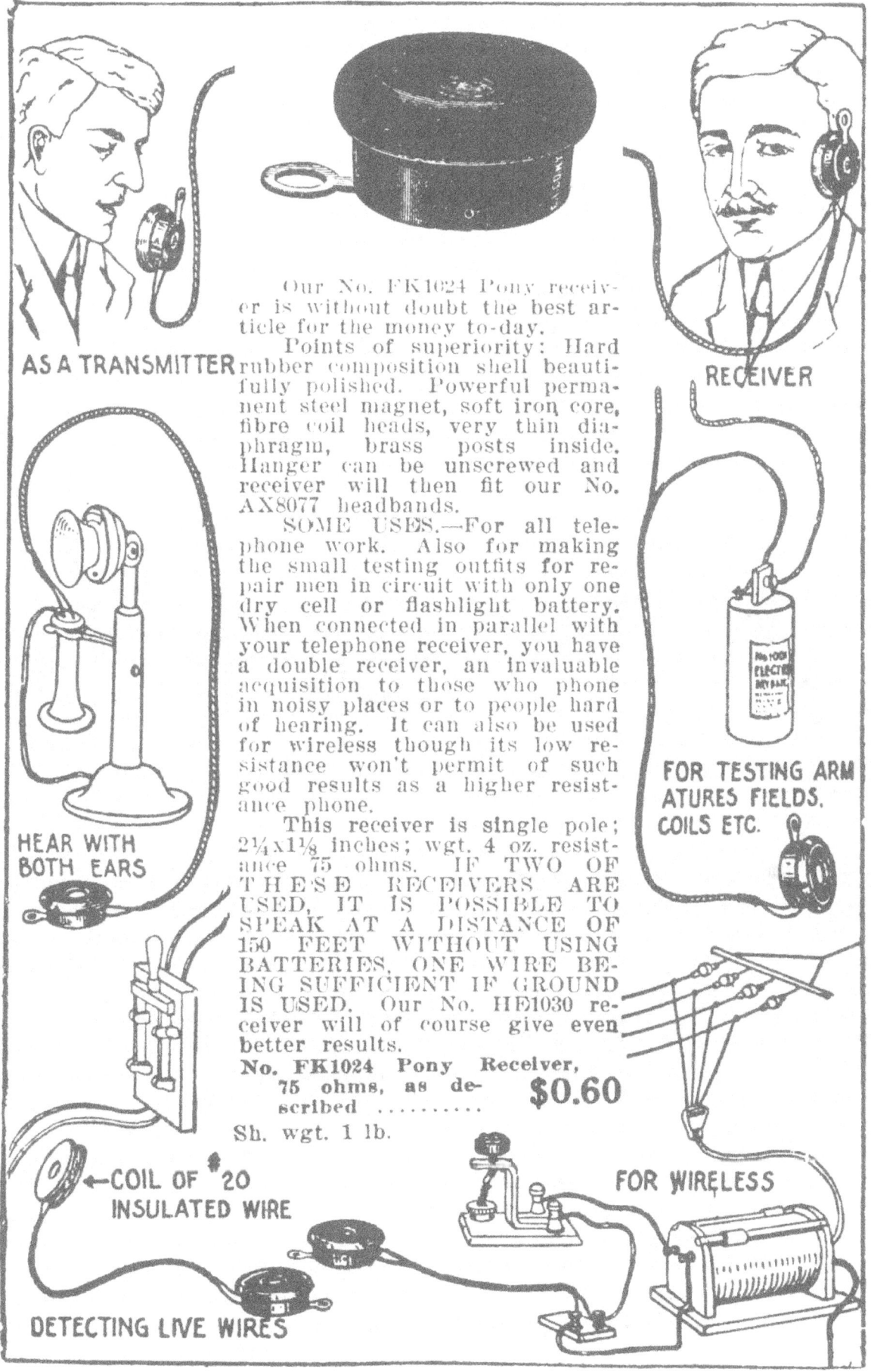

AS A TRANSMITTER

RECEIVER

Our No. FK1024 Pony receiver is without doubt the best article for the money to-day.

Points of superiority: Hard rubber composition shell beautifully polished. Powerful permanent steel magnet, soft iron core, fibre coil heads, very thin diaphragm, brass posts inside. Hanger can be unscrewed and receiver will then fit our No. AX8077 headbands.

SOME USES.—For all telephone work. Also for making the small testing outfits for repair men in circuit with only one dry cell or flashlight battery. When connected in parallel with your telephone receiver, you have a double receiver, an invaluable acquisition to those who phone in noisy places or to people hard of hearing. It can also be used for wireless though its low resistance won't permit of such good results as a higher resistance phone.

This receiver is single pole; $2\frac{1}{4}$x$1\frac{1}{8}$ inches; wgt. 4 oz. resistance 75 ohms. IF TWO OF THESE RECEIVERS ARE USED, IT IS POSSIBLE TO SPEAK AT A DISTANCE OF 150 FEET WITHOUT USING BATTERIES, ONE WIRE BEING SUFFICIENT IF GROUND IS USED. Our No. HE1030 receiver will of course give even better results.

No. FK1024 Pony Receiver, 75 ohms, as described **$0.60**

Sh. wgt. 1 lb.

HEAR WITH BOTH EARS

FOR TESTING ARMATURES FIELDS, COILS ETC.

←COIL OF #20 INSULATED WIRE

DETECTING LIVE WIRES

FOR WIRELESS

Telephone Cords

NO. DE8075

5 FOOT GREEN COTTON BIFURCATED CORD. This cord is used on our No. DX8070 telephones; with 2 nickel tips and 4 loop connections. (See illustration.)
No. DE8075 Each **$0.45**
Shipping weight 2 oz.

No. EE1309 **5 FOOT SILK BIFURCATED CORD,** as used on our FX1305 receivers with 2 tips and 4 loop connections (Shipping weight 2 oz.), each............................. **$0.55**

No. EE6666 Cord is the same in all respects as the No. EE1309. Price the same. Shipping weight 2 oz.

No. AE4005 3 FOOT TELEPHONE CORD with 2 metal tips and 2 loop connections to fit No. AX1024A, AEK8071 Receivers, each. Shipping weight 2 oz...................... **$0.15**

NO. AE4003

AE4003 3 FOOT TELEPHONE CORD, with 4 metal tips, well finished throughout. Each ....
Shipping weight 2 oz. **$0.15**

No. AK6083 **RECEIVER DIAPHRAGM** (Ferrotype). Each.... **$0.10**
Shipping weight 1 oz.

"ANTENIUM" AERIAL WIRES

For a number of years we sold Aluminum wire for aerials, but finally decided to discontinue its sale for several good reasons. To begin with, it is impossible to obtain pure aluminum wire. The commercial aluminum composition wire is notoriously weak and ruptures at 75 lbs., for the No. 14 size. A sharp bend causes it to break almost immediately; it cannot be soldered; it always makes poor contact, on account of its natural oxide film. We had so many complaints on aluminum wire, that we decided to develop an aerial wire that did not have any of the above objections. We finally found it in our present ANTENIUM wire, which not only has none of the objections cited, but has a great many excellent points making it highly desirable for aerials.

ANTENIUM wire is a 30 per cent. copper wire of enormous strength, even surpassing phosphor bronze in strength. Our size wire stands a rupture test of 330 lbs., against 75 lbs. of Aluminum wire. It can be soldered like ordinary copper wire. It can be bent back and forward and is so tough that it cannot be broken, except with difficulty. It makes excellent contact and does not oxidize readily. **It is cheaper than aluminum wire and three times cheaper than copper wire.**

In appearance it is exactly like copper wire, as a matter of fact it cannot be told apart from copper wire.

It has about 50 per cent. less skin resistance than Aluminum wire.

100 feet of our ANTENIUM wire costs $0.45. One pound No. 14 B. & S. Aluminum wire has 200 feet and costs $1.20. Thus ANTENIUM wire is cheaper than the former and is incomparably better. A 600 foot stretch with ANTENIUM wire is an every-day occurrence **and the heaviest sleet will not damage the aerial.** We carry only this one size, which as experience shows is the only kind to use. Heavier wire for aerials is not required as ours is strong enough for the greatest stretch practical.

No. DE9219 **ANTENIUM aerial wire, per hundred (100) feet....** **$0.45**
Not less than 100 feet sold.
Shipping weight per 100 feet, 1 lb.

The "Electro" Rotary Potentiometer

(NON-INDUCTIVE). PATENTED FEB. 28th, 1911

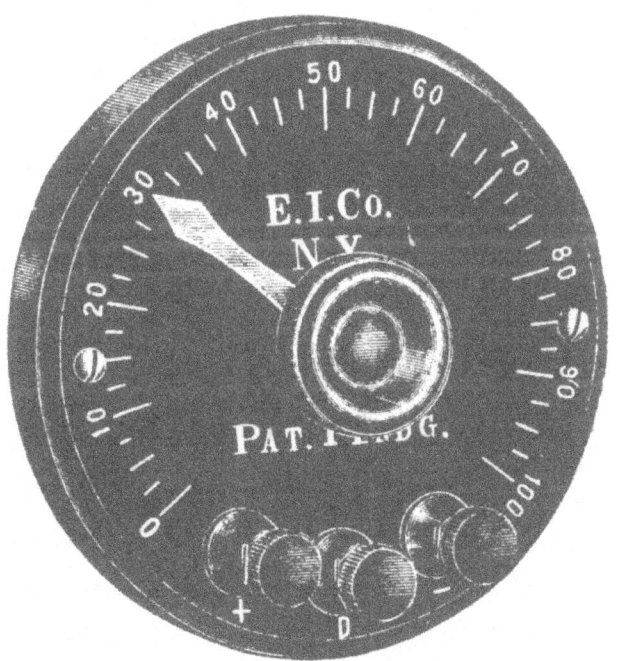

NO. BX9255

There are several unique features incorporated in our instrument which above all takes up a minimum of space being only 4 in. diameter, the thickness of the main body being only ⅝ in.

We use in this instrument a high resistance carbon-graphite rod and the resistance of this instrument is approximately 300 ohms, as experience has taught us that for wireless use only about one hundred to two hundred ohms is generally used, we do not furnish extra rods for this new instrument and 300 ohms will cover all the wants of the operator and experimenter. The most important part is that the movement is rotary and not straight on a long rod, as used in our old style instrument. It will be realized that this is a great advantage, as the rotary movement for wireless instruments comes into force more and more every year. The adjusting knob carries a pointer which moves over an empirical scale which is a great advantage to the operator as he will always know just how much current to give his detector and will easily remember the proper regulation.

All the insulating parts of the instrument are made of molded hard rubber which makes it the most attractive apparatus of this sort ever placed on the market. There is nothing to shrink or warp on this instrument and the construction is beyond criticism. You could not buy a better potentiometer even if you paid ten times the price we are asking for it. The connection is positive. The instrument is always ready and there is nothing to wear out or to be replaced. The carbon-graphite rod is embedded in the hard rubber and it will not break even if the instrument should be dropped. Three of our well-known hard rubber binding posts are provided on this instrument. Also two nickel plated screws to attach the potentiometer to table, wall or an instrument board. The pointer is nickeled and polished. The scale is molded into the hard rubber. The diameter of the rubber thumb handle is 1 inch. An ideal instrument for use with the Radioson detector.

Size over all 4x1½ in.; weight four ounces. WE WILL REFUND YOUR MONEY IF THIS INSTRUMENT IS NOT ALL WE CLAIM FOR IT AND IF IT IS NOT SATISFACTORY IN EVERY RESPECT. Connections are the same as for any potentiometer. Instrument is shipped ready for instant use.

No. BX9255 "Electro" Rotary Potentiometer (patented)........ **$2.00**
Shipping weight 1 lb.

Gentlemen:—   Oakland, Cal.
The circular potentiometer for which I sent HAS ARRIVED O. K. I have it working on my receiving box to-day and IT IS ALL TO THE GOOD. My electrolytic detector is giving GREAT RESULTS; BETTER THAN IT EVER DID BEFORE.

I think that this new potentiometer is the GREATEST THING YOU EVER TURNED OUT.   Yours very truly,   E. W. STONE.

Let us send you free with our compliments lesson No. 9 "The Receiving Apparata" of our famous "WIRELESS COURSE" telling you all about "Potentiometers."

Just attach coupon No. 9 to your order. For information see colored section of this catalog.

Minerals and Crystals

When you buy a mineral or wireless crystal you are interested in only a very few things. First you want to know value. EVERY CRYSTAL SOLD BY US IS TESTED FOR SENSITIVITY. Don't pay more for so-called "special" and "extra" grades. **Our competitors' "special" and "extra" grades are** OUR REGULAR STOCK QUALITY. Now for quantity. Note that we sell by weight wherever possible. When we say you get an ounce, you get an ounce, **not a piece.** This means a big saving to you. Now on delivery. We carry more wireless minerals in stock than any other concern in the world. **We guarantee prompt delivery.** Being the largest buyers and sellers of this class of material we are naturally offered the pick of the world. In that way by buying your Crystals and Minerals from the E. I. Co., you buy the best tested goods that is found at the lowest possible prices. Your first order will convince you of our claims.

BORNITE

Used a great deal abroad. Can be used with a phosphor bronze contact wire, or with zincite. Marvelously sensitive.

No. CE2416 Bornite, per oz. **$0.35**

Shipping weight 2 oz.

GALENA

NO. AE2504

This mineral is thought by many to be one of the most sensitive discovered so far. Used to best advantage by having a fine phosphor bronze or brass wire spring, size about No. 26 B. & S., press **very lightly** on the Galena. We carry only a specially selected cubic crystal grade.

No. AE2504 Galena, per ounce **$0.15**

Shipping weight 2 oz.

COPPER PYRITES

Very sensitive and very stable. Even sensitiveness along whole surface. Not easily jarred out. Use phosphor bronze contact wire. GUARANTEED 100 PER CENT. PURE.

No. CE2419 Copper Pyrites, per oz........ **$0.35**

Shipping weight 2 oz.

ZINCITE

The aristocrat of all wireless minerals. Too well known and too far famed to praise it here. Undoubtedly the most sensitive of all crystals. GUARANTEED 100 PER CENT. PURE.

No. ABE2417 Zincite, per oz............... **$1.25**

Shipping weight 2 oz.

No. CE2418 Zincite, ¼ oz. **$0.35**

Shipping weight 1 oz.

SILICON

There are two kinds of this material: Silicon crystals and fused Silicon. The former, manufactured in this country, is absolutely unfit to use; the latter, imported by us, is the only kind that should be used. It comes in chunks and somewhat resembles graphite. It is very hard and extremely brittle.

No. CE9209 Silicon, per oz. **$0.35**

Shipping weight 2 oz.

No. AE9209a Silicon, ¼ oz. **$0.15**

Shipping weight 1 oz.

IRON PYRITES

Our iron pyrites is all imported Spanish stock that may be used for years without deterioration. Very sensitive.

No. CK2505 Iron Pyrites (Ferron), extremely sensitive, per oz. **$0.30**

Shipping weight 2 oz.

Minerals and Crystals
(Continued)

MOLYBDENITE

This new substance is the only one discovered so far which does not get out of adjustment, when used in a sensitive Detector, and when placed near a sending gap. Most substitutes suffer a great deal from strong sending currents, but it is impossible to damage the adjustment of the Molybdenite Detector, and a heavy discharge does not affect it. Molybdenite proves quite sensitive when distant stations are to be picked up.

No. EK9210 Molyb-denite, per oz...... **$0.50**
Shipping weight 2 oz.

CARBORUNDUM

Specially selected for experimenting with the Carborundum Detector. Quite sensitive. Used by commercial companies for many years.

No. CK9308 Carborundum, per oz........ **$0.30**
Shipping weight 2 oz.

PEROXIDE OF LEAD

No. CK2506 Peroxide of Lead, compressed tablets, each **$0.30**
Shipping weight 2 oz.

MINERAL SETS

No. GK2502 Zincite and Copper Pyrites (Perikon), per set **$0.70**
Shipping weight per set 4 oz.

MINERAL ASSORTMENT

Consisting of generous pieces of each of the nine minerals and crystals shown on these pages. An excellent assortment for the wireless experimenter. Each mineral in a separate box. No Radiocite supplied.

No. AEK2346 Mineral Assortment (9 minerals) **$1.50**
Shipping weight 1 lb.

Wireless Code Chart

This code chart has been brought out by us pursuant to a large demand by our enthusiastic wireless friends, who like to have the three codes, the Morse, Continental and Navy, before their eyes when sending or receiving messages. This is truly a beautiful chart, being arranged in such a manner that a letter or figure can be "spotted" instantly, without the eye searching for precious seconds. The dots and dashes are very heavy and large and can easily be read 10 feet off. There are in addition, a list of abbreviated numerals as used by Continental operators; also the usual wireless abbreviations used by most of the fraternity.

The chart measures 9x11 inches and is printed on stiff cardboard. It will make a fine addition to any wireless station and it will make the latter look businesslike.

Comes in black on white background only.

The latest feature of this article is that on the back we now have the International Morse Code and conventional signals, also the list of abbreviations to be used in radio communication and as adopted by the International Radiotelegraphic Convention. Room is also left for a private code if desired.

No. AK2501 Wireless Code Chart.... **$0.10**
By mail, extra $0.03.

10c

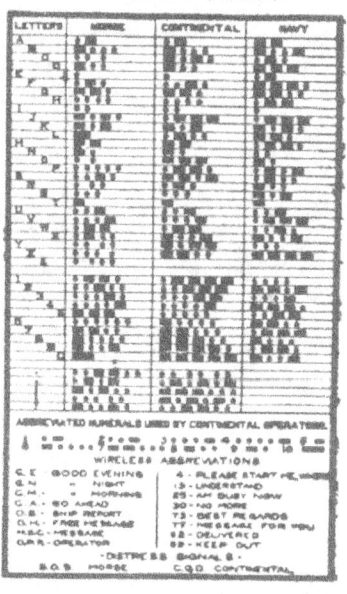

No. AK2501

The "Electro" Lilliput Buzzer

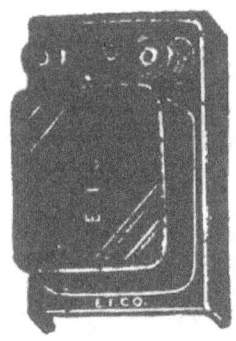

NO. EE965

These buzzers represent the latest advance in all metal buzzers. The cover and base are made of stamped cold rolled steel. It has a pivoted armature. The tension is easily altered, and the contact adjustment held securely by a spring nut. The ribbed edges of the cover spring tightly over the base, making it likewise readily removed, allowing inspection and adjustments of the moving parts if desired. Positively dust and insect proof. This is just the buzzer to use for testing the minerals of the receiving wireless set, as it is compact and very neat. Wound to 3 ohms. Base and cover are finely japanned. Size over all 2⅛x1 5/16x⅝ in.

No. EE965   The "Electro Lilliput" Buzzer...... **$0.55**
Shipping weight 4 oz.

Tin Foils

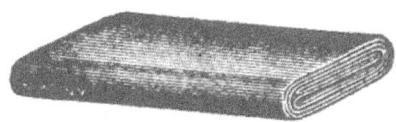

NO. CE4335

TIN FOIL. We have 2 grades of tin foil listed below. For small paper condensers we recommend our No. CE4335. Our No. CK6252 is intended for very heavy work, such as large capacity condensers up to 5 K. W. Not less than 1 lb. of a size sold.

Catalogue No.	Thicknesses	No. of Sq. Inches to lb.	Remarks	Price per lb
CE4335	Thin	1728	Suitable for paper condensers.	$0.35
CK6252	Heavy	600	For large transmitting condensers.	$0.30

Shipping weight per lb., any style, 2 lbs.

Solderall

NO. DK1146

SOLDERALL is a wonderful solder and non-corrosive flux combined, in paste form, and contained in a collapsible tube, always ready for instant use. No acids, rosin or flux necessary.

All you have to do is to unscrew the cap from the nozzle of the tube, squeeze a little SOLDERALL on the parts to be soldered, and heat with a match, hot iron or torch, and the work is done.

Large holes can be soldered (something impossible to do with other solders) by building a pyramid of SOLDERALL over the hole and then applying a match or torch at short intervals, so as to melt slowly.

Indispensable in the Home, Shop, Garage, Laboratories, for repairing Kitchen Utensils, Toys, Leaky Pipes, Tin Roofs, Automobile Parts, instruments, Models Etc., Etc.

Will be found of the greatest use by Electricians, Telegraph and Telephone Linemen, Plumbers, Gas Fitters, Automobilists, Motor Cyclists, Dentists, Physicians, Jewelers, Engineers, Campers and Sportsman. Size of tube ¾x3¼ in.

No. DK1146 SOLDERALL (Shipping weight 4 oz.) Per tube...... **$0.40**

Gentlemen:—   New Castle, N. B.
    I purchased one of your No. 1043 Lamps from a friend ABOUT EIGHT MONTHS AGO AND HAVE USED IT VERY CONSTANTLY. IT WAS VERY SATISFACTORY and this is partly the reason WHY I AM NOW ORDERING four lights instead of one; AS I FEEL CONFIDENT THAT THEY WILL BE ENTIRELY SATISFACTORY. Also I will be glad TO RECOMMEND YOUR GOODS FOR QUALITY AS WELL AS PRICE.   Yours respectfully,   R. D. W. FLEWELLING.

The "Electro" Precision Coherer

This coherer is used where extreme accuracy is desired on all distances up to 30 miles. The silver plated, amalgamated coherer plugs are fitted with micrometer screws, which allow the screws to move forward or backward. The regulation is so correct that the plugs can be moved less than 5/1000 inch at a time. All parts are made of burnished brass highly finished. The base is mahogany. No. BIE1295 Coherer and Decoherer is made adjustable, so that the strength of the tapper can be regulated. Size over all 7¾ x 4¼ x 2 ins.

No. BIE1295

No. BIE1295 **Coherer and Decoherer, complete**................ **$2.95**

Shipping weight 2 lbs.

High Capacity Condensers

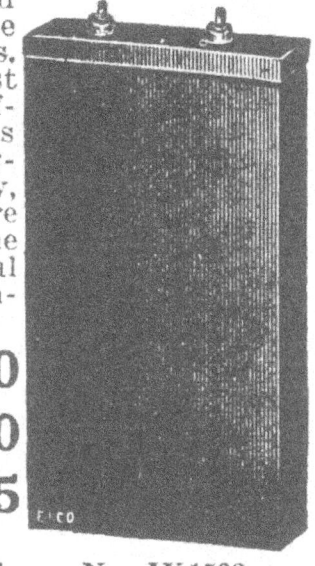

Few articles are so hard to make and get good results from, yet are so important an article to the electrical experimenter, as high capacity condensers. Those listed by us here are made of the very best grade of rice paper and tinfoil impregnated with paraffine. Capacities, while high, are ideal for telephones or ringing circuits, for experiments with duplex telegraphy, artificial cable capacities, wireless telegraphy, and as spark coil condensers, etc., or any service where high capacity, low tension condenser can be used. The terminals are brought out in a neat and substantial manner, and clips are supplied for mounting the condensers either singly or in pairs.

No. FK1582 **High Capacity Condensers** (½ microfarad) . Size 4½ x 1¾ x ½ in........ **$0.60**

No. IK1583 **High Capacity Condensers** (1 microfarad). Size 4 x 2⅝ x ½.............. **$0.90**

No. ABE1584 **High Capacity Condensers** (2 microfarads). Size 4⅜ x 2¼ x ¾............. **$1.25**

Shipping weight 2 lbs., any capacity

These Condensers cannot be used for high tension work

No. IK1583

The "Electro" Watch Case Buzzer

The highest grade of material and workmanship has been incorporated in the "Electro" watch case buzzer illustrated in our engraving. The buzzer has a solid brass case, beautifully and durably nickel-plated. The springs are made of phosphor bronze, with pure silver contacts, assuring long life to the buzzer. The insulation is perfect.

This buzzer is unequalled for a portable testing instrument, as it may be carried in the pocket without inconvenience. It is suitable for any work requiring a compact and neat buzzer. The sound is pleasant, and as clear as that obtained from buzzers of larger dimensions. Size over all 1⅝ x ½ in.

No. GE950

No. GE950 **"Electro" Watch Case Buzzer** **$0.75**

Shipping weight 4 oz.

"Electrite" Specialties

Parts listed below are made of our new secret composition "Electrite." This material has highest electrical value. Has all the characteristic properties of hard rubber, including jet black color and high polish, but greater strength than rubber. **ALL CUTS FULL SIZE.** This material is not affected by water or acids.

Electrite Knob. Is 7/16 in. diameter and ½ in. long, with 8/32 threaded brass bushing.
No. 809 Electrite Knob, each.. **$0.05**
Shipping weight 4 oz. per doz.

No. 809

TYPEWRITER KNOB

While our Thumb Screw No. 6011 has been found very useful in the construction of many instruments, we have had a persistent demand for a larger handle with a sleeve and a convenient hole and tightening screw. We meet this demand with the special knob illustrated. It may be used for a great variety of purposes. The dimensions are: Diameter of knurled head, 1½ in.; length of sleeve, about 7/16 in.; diameter of sleeve, ⅜ in.; diameter of hole in sleeve ¼ in. A 6/32 screw holds the metal rod in place when in the hole of the sleeve. Shank is nickel plated.

No. 6013

No. 6013   Typewriter Knob **$0.25**
Shipping weight 4 oz.

SWITCH HANDLE

A favorite style with many constructors owing to its very neat appearance and knurled sides. Is ⅞ in. long and has a threaded insert for 6/32 screw. Is ¾ in. diameter.
No. 945 Electrite Switch Handle, each **$0.05**
Shipping weight 4 oz. per doz.

No. 945

ELECTRITE HANDLE

This handle may be used for a number of purposes besides a switch handle. It is of suitable size for handles on spark gaps up to 3 KW. Size of handle: Body 1½ in. Screw projects 7/16 in. This cut is not full size.
No. 6841   Price, each **$0.15**
Shipping weight 2 oz.

No. 6841

ELECTRITE THUMB SCREW

This Thumb Screw is useful for constructing many kinds of instruments. It is a favorite with wireless experimenters who use this part in their detectors. Equipped with pointed screw ⅞ in. long Thread 8/32. Diameter of head 1 in. Note—This cut is not full size.

No. 6011

No. 6011   Thumb Screw, Price each **$0.20**
Shipping weight 2 oz.

ELECTRITE TELEGRAPH KNOB.

Used on keys, etc. 1⅛ in. diameter. Has 8/32 screw molded in. Screw projects ⅛ in.
No. 908 Electrite Telegraph Knob, each **$0.10**

No. 908

Shipping weight 4 oz. per doz.

THE "ELECTRO" SWITCH HANDLE

A complete switch lever consisting of a 1 in. Hard Rubber Typewriter knob with nickel plated phosphor bronze blade 1¼ in. long.

No. 9790

This switch lever can never work loose nor make poor contact as it is kept tight by a forked spring fastened from the bottom, as shown in cut. An ideal switch for loose couplers, inductive tuners, switch boards and innumerable uses that suggest themselves to the constructor. It is the only switch which can never become loosened, no matter how much it is worked.
Complete mechanism consists of seven parts as shown.
No. 9790 "Electro" Switch Handle, as described, complete (Shipping weight 4 oz.) **$0.30**

KEY KNOB

Electrite Knob. Is ⅝ in. diameter and ½ in. high with 8/32 threaded brass bushing.

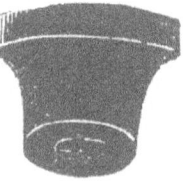

No. 507

No. 507   Electrite Knob. each **$0.05**
Shipping weight 4 oz. per doz.

ANTENNA INSULATORS
The "Electro" Junior Strain Insulator

For small aerials, used principally for receiving, a small but highly efficient insulator is desirable. In view of the fact that our Ball Antenna Insulators appear to some to be too large for this purpose the present insulator has been developed.

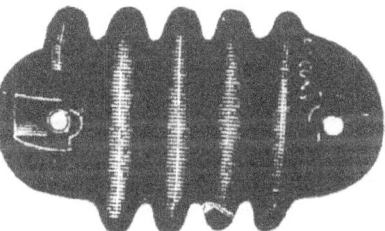

NO. AB9461

It consists of a small but heavily and deeply ribbed brown glazed porcelain, whose surface area, due to the thread is twice as great as it would be if it were perfectly smooth. In other words it is as efficient as a cylindrical insulator of twice its length and more than twice its weight. It has a protected and smoothly turned hole in each end for wires.

Another advantage of this type of insulator over the plain ridges or corrugated insulator is that if used at an angle it conducts any rain or water down to a drip point, at the lowest end rather than holding it in a puddle between the ridges, which reduces the insulating qualities.

Size over all 2¾ in., diameter 1¼ in., net weight each 3½ oz.

No. AB9461 "Electro" Junior Strain Insulator.................. **$0.12**
Shipping weight 3 lbs. per eight.

The "Electro" Ball Antenna Insulator

The size of this Insulator over all is 3¼x2½ inches. Weight 7½ ounces. The insulator is made entirely of porcelain in one piece and has a triple coating of brown glaze. The insulating value is of the highest order and greater than similar insulators. It will hold 35,000 volts.

All the grooves are undershot and this feature is responsible for the fact that the new insulator "sheds the water like a duck."

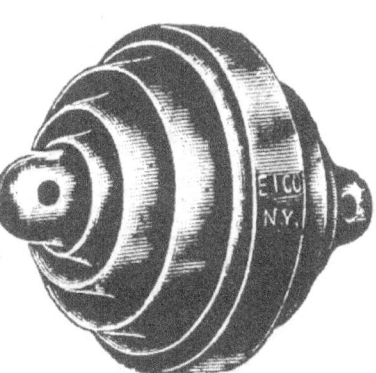

NO. AF10007

We guarantee that you will be pleased with this insulator. It is the cheapest high-grade insulator ever placed on the American market and stands in a class by itself.

We recommend one insulator on each end of an aerial strand for receiving. For sending there should always be two of the insulators in tandem; this will afford sufficient insulation up to 1 K. W. transformer.

No. AF10007 "Electro" Ball Insulator, as described, each........ **$0.16**
Shipping weight 1 lb.

The "Electrose" Ball Antenna Insulator

The insulator which we offer herewith is made of moulded electrose composition which is acknowledged to be one of the best insulators on the market. The eye hooks are moulded right into the electrose and the insulator will stand a strain of 600 to 700 lbs. Will stand 45,000 volts. For sending it will hold the discharge of a 4-inch coil and therefore can be used in connection with small sending outfits without trouble. The eye rings are wrought iron. Size over all is 3½x2½ inches.

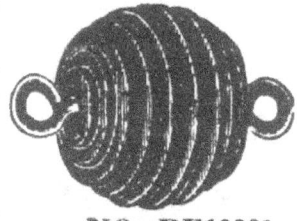

NO. BE10001

No. BE10001 Electrose Ball Antenna Insulator, as described. **$0.25**
Shipping weight 1 lb. Each.............

Commercial Wireless Insulators

NO. FK10002

These insulators are used on large aerials and are built very substantially in order to stand the enormous strains sometimes experienced in heavy storms. These insulators are now used by the UNITED STATES GOVERNMENT and are made of moulded "Electrose." They will stand discharges of 80,000 volts. Deep corrugations are provided to reduce surface leakage. Powerful wrought-iron rings are imbedded at each end.

Size: 10½ inches long, 1½ inches in diameter.

No. FK10002 Commercial Wireless Insulator, as described, each **$0.60**
Shipping weight 2 lbs.

The "Electro" Wireless Lightning Switches

100 AMPERE

NO. BEK1616

The Underwriters' Rules in all cities now call for lightning switches, wherever aerials are erected on top of buildings.

The rules prescribe either 250 or 600 volt, single pole, double throw switches, which must be fastened outside of the building. A No. 4 B. & S. wire is specified in all cases, to run from the aerial to the switch and thence to the ground, on the outside of the building.

The aerial should always be grounded when not in use, to protect the house from lightning. Connecting diagram is given herewith. Our switches are of the standard type. All metal parts are of pure copper, base is slate. Strong, durable handle is furnished.

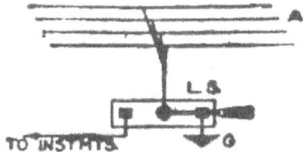

No. BX1616 measures 14x2½x3 inches over all. Its carrying capacity is 100 amperes.

No. BHE1617 measures 17x3x3 inches over all. Its carrying capacity is 100 amperes. **DON'T BUY 60 AMPERE SWITCHES.**

No. BEK1616 250 Volt Lightning Switch. Shipping weight 6 lbs... **$3.00**

No. CCE1617 600 Volt Lightning Switch. Shipping weight 7 lbs. **$3.50**

Ground Wire for Lightning Switches

(Approved by the Underwriters.)

This wire has a soft iron core while the outside is pure SOLID copper. The copper forms about one-third of the entire wire. Only one size carried, No. 4, which has 9 feet per lb.

No. F4004 Size 4 B. & S. soft copper clad wire, per foot.......... **$0.06**
Shipping weight 2 lbs. for each 10 feet.

Carbon Grain Transmitter

This is a special design of transmitter for long distance work. It may be used with satisfaction on wireless telephone sets where a heavy current is to be passed through it. This is a first-class instrument in many respects.

A telephone transmitter has only one function to perform and it either does that right or the most expensive telephone equipment is useless. Same applies especially when a transmitter is desired for experimental purposes. Our transmitter has a very low resistance and a high current capacity. Above all it don't transmit that tinny sound as cheap transmitters do. Altogether a finely nickel plated article at a very low price. Size over all 3¼x2½.

No. BX6080

No. BX6080 Carbon Grain Transmitter, each.. **$2.00**
Shipping weight 1 lb.

The "Electro" Antenna Switch

As illustration shows this is a three-pole, double throw switch. As will be seen the throw to change the switch over is only about 1 inch, making it almost instantly. The two end blades are at an angle of 140 degrees and the construction of this switch is unlike others. By referring to the diagram it will be seen that when the switch is thrown for receiving the primary of the coil is disconnected. If accidentally the sending key should be touched it will be impossible to damage the receiving instrument, as the coil can under no circumstances operate. The diagram shown is standard, but of course many other connections can be devised by the experimenter. All metal parts are pure copper.

Hard rubber handle is provided as switch handle. This switch will stand the discharge of a 4-inch coil without jumping across. It can be used in connection with a transformer up to 5 K.W. All copper parts are very heavy.

The switch can be screwed down on any table or wall. Size of base 7x7 inches, height over all 4 inches, when lever is down; when lever is up, height is 5 inches. There is at the present no quicker wireless throw switch on the market.

NO. BX8100

No. BX8100 "Electro" Antenna Switch, as described, price.... **$3.00**
Shipping weight 3 lbs.

The "Electro" Zinc Spark Gap

We have placed on the market a good many articles during the past, but we pride ourselves that our little "Electro" Zinc Spark Gap, for efficiency, neatness, simplicity and low price, stands unequaled.

While our old style jump spark balls were well suited for short distances, for which purpose they were unmatched, the "Electro" Zinc Spark Gap

No. GE9220

is intended to do real hard work—even commercially for short distances.

The peculiar properties of a small zinc spark gap make it particularly efficient for sending, especially when a sending condenser is used.

Any size spark coil up to 6 inches can be used successfully. It has been found in the past that if zinc is used in the sending spark gap, same will transmit fully twice as far as brass or any other metal, hence, as usual, we use the best. If a single small Leyden jar is shunted across the gap and if, for instance, a 1-inch coil is used, an intense blue mass of fire will crash across the gap with a roar—exactly as you hear it in the large commercial and government stations. If you never saw our Zinc Gap in operation, you will hardly realize its power. Besides, it may also be used as an ANCHOR GAP in the antennae, which serves as an automatic switch. The "Electro" Zinc Spark Gap has two zinc rods 3-16 inch diameter, and 2½ inches long, having a hard rubber handle at each end, making it possible to adjust the gap while sending. Stands which are finely plated are mounted on heavy hard rubber composition base

Size of base, 2½ x 3½ in. Size over all, 6 in. long, 2 in. high. **$0.75**
No. GE9220 "Electro" Zinc Spark Gap, as described.............
Shipping weight 1 lb.

The "Electro" Rotary Spark Gap

NO. ADX2382

The advantages of the rotary spark gap are too well known to require much comment. Sure it is, however, that the "Electro" Rotary Spark Gap will give you an efficient, high pitched spark that will increase your sending range at least 30 per cent. (and probably more), besides making your emitted signals more easily read. The motors are all standard stock motors that have been manufactured for years,

are well constructed, operate at high speed and are perfectly dependable under all conditions. The disc is of solid Bakelite, 4¾ in. in diameter and with 12 large zinc gap contacts that have been carefully turned and ground to size. The disc runs perfectly true. The capacity of the gap is 1 K.W. and this capacity can be carried continuously. Base is of **ELECTRITE** that can't leak or crack, as does slate and marble. Binding posts are nickel plated and very conveniently placed. The stationary electrodes are of zinc and fully adjustable to take up wear and burning of contacts. **All contacts are renewable by use of pliers and screw driver only.** Owing to its high speed the "Electro" Rotary Spark gap is especially valuable for use in connection with Tesla Transformers and high frequency outfits. You positively cannot buy a better rotary spark gap at any price.

Size, base, 7x9 in.

No. ABX2382 "Electro" Rotary Spark Gap, as described, with 6 volt battery motor **$12.00**
Shipping weight 18 lbs.

No. ADX2382 D.C. "Electro" Rotary Spark Gap, with 110 volt D.C. motor **$14.00**
Shipping weight 20 lbs.

No. ADX2382 A.C. "Electro" Rotary Spark Gap, with 110 volt 60 cycle A.C. motor **$14.00**
Shipping weight 20 lbs.

Lincoln, Nebr.

Dear Sirs:—
Last year I ordered a number of things of you, among which was an "Electrolytic" Interrupter for my wireless station. I have used it very hard all the while except during the summer. I found that it worked just as well after three months of non-use as it did when I got it. The "Telegraph Key" I received of you is a dandy, as is the Ball Bearing slider.
ASHLEY WILLIAMS.

When ordering one of our Spark Gaps, Telegraph, or Wireless Keys, permit us to send you FREE with our compliments, lesson No. 5 "The Amateur Transmitting Sets and Apparata" or lesson No. 6 "Transmitting Sets" or lesson No. 7 "New Transmitting Systems" or lesson No. 15 "Learning to Operate" of our famous "WIRELESS COURSE." More practical knowledge is contained in these lessons than in big books.
Just attach one or all coupons Nos. 5, 6, 7 or 15 to your order. For further information see colored section of this catalog.

The "Electro" Telegraph Keys

A new departure in telegraph keys. There has long been a demand for a good, efficient, but cheap telegraph key and the one which we are now manufacturing complies with all demands that anyone could possibly make of a low price key. The parts are mounted on a solid hard rubber composition base, size 2½x3½ inches, ¼ inch thick. All metal parts are nickel plated and polished and the contact arrangement is simple but absolutely sure. A standard telegraph knob one inch in diameter in hard rubber composition is furnished. The No. CE1118 Key has two of our standard rubber binding posts, while the No. DK1119 has three of them. This key works easily and there is nothing to get out of order. It will make a handsome addition to any instrument table.

No. DK1119

No. CE1118 Single Circuit "Electro" Telegraph Key composition base, as described. Shipping weight 1 lb........... **$0.35**

No. DK1119 Double Circuit (Morse) "Electro" Telegraph Key composition base, as described. Shipping weight 1 lb. **$0.40**

The "Electro" Telegraph Key

These steel lever, standard telegraph keys are a radical departure from the old style metal keys and the amateur as well as the professional will find these keys far superior to anything that has been offered heretofore. Our new departure is centered in the fact that instead of using a metal frame, which is so liable to short circuit the different parts, we use a ½ in. insulated base. This not alone gives the instrument a classy appearance, but it enhances at the same time the insulation a great deal, and our keys to-day are without doubt the most beautiful and the best built on the market, barring none. We do not use a spiral spring to operate the key but use a special kind of a tongue spring which works a great deal easier and smoother than the old style spring. The lever is solid steel, nickel plated and highly polished as are all the metal parts on the key. There is absolutely no lateral motion and the trunnions cannot possibly get loose. The contacts are pure silver. A generous hard rubber handle and two large binding posts are furnished. If you have once used this key you will never use another.

No. ABE1117

We only furnish one style of this key, namely, the one with top connection. No leg connections are furnished.

Size over all 6½x2¾x1¾ in. Size of base ½x2½x4¼ in.

No. ABE1117 Steel Lever Key with Insulated Base as described, each. Shipping weight 1 lb........................ **$1.25**

Do you know that we will send you FREE, with our compliments the following lessons of our famous "WIRELESS COURSE" (or any other lesson you may choose) when ordering our Sending Helices, Oscillation Transformers, Sending Condensers, Leyden Jars, etc.:

Lesson No. 1 "The Principles of Electricity" or lesson No. 4 "The Principles of Wireless Telegraphy" or lesson No. 5 "The Amateur Transmitting Sets and Apparata" or lessons Nos. 12 and 13 "The Hook-Ups and Connections."

Just attach the corresponding coupons to your order. For further information consult colored section of this catalog.

The "Electro" Wireless Key

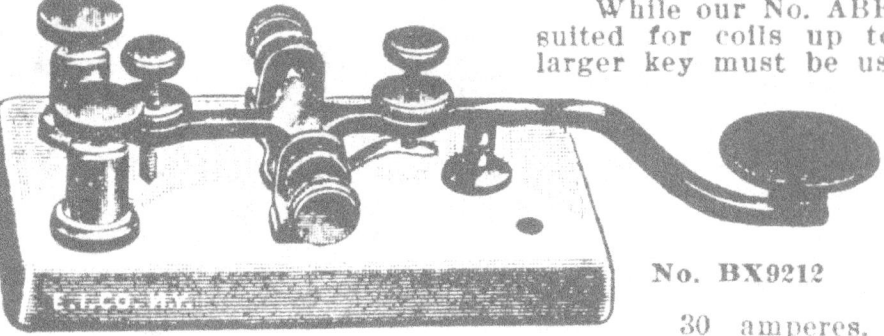

30 AMPERES CAPACITY

While our No. ABE1117 Key is well suited for coils up to 2 in. spark a larger key must be used for the more powerful coils, from 3 to 12 in. spark length, also in connection with our No. GGE8050 transformer or coil using up to 30 amperes. Our key will positively not heat up even if 30 amperes are used for hours at a time. This key is similar to our No. ABE1117 except that it is very much heavier and two extremely large binding posts that can take a No. 6 wire, are furnished. The contact points are solid silver and measure ⅜ in. in diameter, ⅛ in. thick; they are built in such a manner that they can be exchanged in less than two minutes. **No tools being necessary.**

As a **HEAT PROOF** Insulated Base is used, it will be understood that the insulation is the best that can possibly be had and there need not be any fear of short circuit as with metal base keys. For the price at which this key is sold it is positively the greatest bargain offered in wireless keys to-day. We will immediately refund the purchase price if the key is not absolutely satisfactory in all respects. All metal parts are highly nickel plated, hand polished and buffed. Size of base ½x2½x4¼ in. Size over all 6½x2¾x2 in.

No. BX9212 Wireless Key with Insulated Base, as described.... **$2.00**
Shipping weight 2 lbs.
No. CK9213 Upper Contact for above key....................... **$0.30**
Shipping weight 2 oz.
No. CK9214 Lower Contact for above key....................... **$0.30**
Shipping weight 2 oz.

NO. BBE1718

The "Electro" Kick-Back Preventer

All transmitting sets in Wireless stations, employing commercial light or power circuits for the source of energy, are required to properly protect the circuit against Kick-backs from the spark coil or transformer. To this end, the Fire Underwriters require that two, one-half micro-farad, fixed condensers, be connected in series across the primary circuit, supplying the transformer or spark coil. The centre connection between the two condensers is to be grounded to a good damp ground connection, as in cut, or to a water pipe, on the street side of all meters, etc. The ground wire should be run on insulators, and be of the same size as the primary leads of the transmitting set. The proper capacity condenser has been

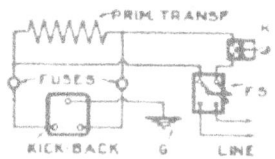

developed by us, and is very compact and efficient. It is made of heavy tin foil and a good dielectric; the enclosing case being of glass. The condensers are then sealed in a high grade sealing compound giving a superb insulation, that cannot be surpassed. Get one of these condensers to-day, and have your station protected according to the Underwriter's rules, before you get into trouble.

Size, 3¾ x 4 x 6 inches. Shipping weight 8 lbs.
No. BBE1718 Kick-back Preventer, as described. Price........... **$2.25**

The "Electro" Hot Wire Ammeter

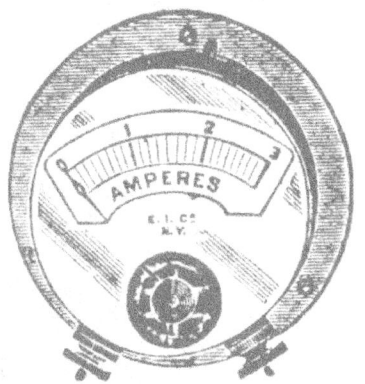

No wireless sending station can possibly operate at its highest efficiency unless it is adjusted to radiate the maximum of current.

There is only one way to determine how much current is being radiated and that is by placing a hot wire ammeter in series with the aerial lead-in. This fact is known by many amateurs who really want a Hot Wire Ammeter but who were kept from buying one on account of the former high price of these meters on the market. Knowing the demand for a good but reasonably priced hot wire ammeter we put our engineers on the problem, and they, working in conjunction with our skillful French meter-mechanics succeeded in producing a meter more accurate, **better looking and more substantial and at a lower price than any hot wire ammeter heretofore offered.**

This meter is the greatest bargain at its price that we have ever offered the Electrical Experimenter; you are absolutely safe in buying the "Electro" Hot Wire Ammeter for **it is sold with a money back guarantee if it is not as represented.**

Specifications:—Case 3 in. diameter, 1⅜ in. high, nickel plated, reading 0-3 amperes, has temperature compensating lever to keep pointer at zero in any climate. Pointer brought back to zero when in use by perfect spring mechanism. Binding posts are large, perfectly insulated and conveniently placed. Bearings are jeweled and pivots are of hardened steel ground to size by hand.

Get this meter now and make sure your station is operating at maximum efficiency.

No. CEK9100 "Electro" Hot Wire Ammeter reading 0-3 amp...... **$3.50**

Shipping weight 3 lbs.

NO. CEK9100

The "Electro" Leyden Jars

Our jars embody the best workmanship. The glass used is guaranteed to be the thinnest hard glass, free of all harmful salts. We could turn out these jars at almost half the price listed, by using lead glass, but such jars will leak badly and soon crack. Our jars may be subjected to a remarkably high potential and are very hard to puncture. The tinfoil is at least twice as heavy as that used in other jars and will not "blister."

Solid brass top binding posts are used and the glass not covered with tinfoil, inside and outside, is finished in black enamel, giving the jar a very beautiful appearance, besides preventing leakage. All our jars can be charged with even our ½-inch coil and the discharging crash of even our 1-pint jar can be heard a long distance. It is powerful enough to kill a cat with a single discharge. These jars cannot be charged by a static machine but are for use on coils and transformers only. The following sizes are made:

No. IE9221 Leyden Jar, as described, 1 pint........ **$0.95**
Size over all 3x8 in.
Shipping weight 2 lbs.

No. ABE9222 Leyden Jar, as described, 1½ pint.... **$1.25**
Size over all 3x8½ in.
Shipping weight 3 lbs.

No. AGE9223 Leyden Jar, as described, 1 quart.. **$1.75**
Size over all 4½x9 in.
Shipping weight 4 lbs.

No. BX 9224 Leyden Jar, as described, 2 quarts.. **$2.00**
Size over all 5x11½ in.
Shipping weight 5 lbs.

No. IE9221

The "Electro" Commercial Oscillation Transformer

This instrument fills a long felt want among amateurs and has been designed to be used with power up to 1 K.W.

Now that the Wireless situation is clear and we know that the amateur may use a wave length up to 200 meters, we thought we could do no greater favor to the American amateur than build an apparatus which would confine the wave length to 200 meters, and at the same time obtain

$6.00

Weight
15 Lbs.

No. FX9600

high efficiency. In connection with a four-wire aerial, total length of fifty feet with any transformer up to 1 K.W., a wave length of not more than 200 meters will be sent out.

The adjustment of this apparatus is the most complete that could be thought of. There are two spirals of heavy Metal Wire, each spiral having eight turns. The two spirals can be separated by moving the spiral on the right back and forward, the maximum separation being 10 inches.

We use no helix clips, but the adjustment is made by means of a slider mounted directly on the back of each spiral. By means of these sliders—adjustments which vary the inductance to a half turn, are readily accomplished. This feature cannot but recommend itself and has never been attempted before in any other similar instrument. By means of the handle 4, the movable spiral can be adjusted back and forth, which assures any adjustment desired. This apparatus is especially recommended for close

The "Electro" Commercial Oscillation Transformer

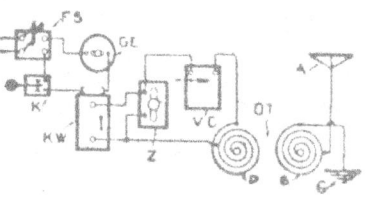

tuning and is invaluable for **Wireless Telephony** where it positively stands unequalled.

The construction and workmanship on the apparatus is of the highest order and stands distinctly by itself. The wood is cherry throughout, mahogany finish, hand rubbed polish, ⅝ inch thick. The Oscillation Transformer is shipped flat and takes up but little space when shipping. The full size of the instrument is 16x14x12 inches. All metal parts are brass nickel plated, except the Aluminum Wire.

There are six generous nickel binding posts. The Metal Wire spirals are fastened by a unique process, never attempted before. The whole instrument will make a valuable addition to any wireless station. We positively guarantee the working efficiency and wave length of this instrument and will cheerfully refund the purchase price if not entirely satisfactory and equal to our description.

No. FX9600 "Electro" Commercial Oscillation Transformer as described. Shipping weight 15 lbs. **$6.00**

The "Electro" Sending Helix

LAMP FOR PILOT DEVICE SOLD EXTRA　　　**WITH PILOT LAMP DEVICE**

This is something new in helices, and although we have been making similar helices for years, we have improved our old types in several ways. The new DX8271 Helix is built entirely of SOLID MAHOGANY throughout (no imitation mahogany). There are eight turns of heavy Aluminum wire, two fine helix clips with several feet of best imported high tension cable. Hard rubber binding posts for the two other connections are provided.

The new departure, however, is found in the PILOT LAMP device. This unique feature is original with us and has never been offered by any other concern. "AS USUAL WE LEAD." The pilot lamp has NO METALLIC CONNECTION with the

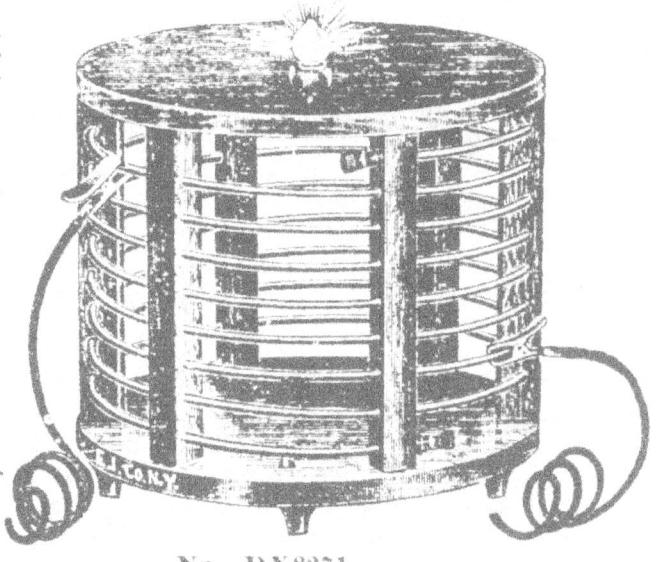

No. DX8271

helix, as the lamp lights only by the inductive effect of the helix. A loop of wire fastened under the top board connects directly with the lamp. This loop is fully 2 in. away from the helix spiral. Instead of using a hot wire ammeter, the pilot lamp is used, and when the lamp lights up brightest you know that you are radiating the maximum amount of energy.

We do not furnish a lamp with the helix, as any small incandescent lamp, Carbon or Tungsten, may be used. We furnish a socket which takes ONLY miniature base lamps. A lamp is not furnished as for each different coil, a different voltage lamp is used. With a 2-inch coil use a 2-4 volt lamp. With a 3-inch coil or ¼ K.W. coil or transformer use a 4-6 volt lamp. With a ½ inch K.W. transformer use a 10-16 volt lamp. Inasmuch as all depends on the coil or transformer used, each helix should be tried with various voltage lamps till the right lamp is found. Sizes are: 10 in. diameter, 9 in. high; thickness of wood is ½ in.

No. DX8271 "Electro" Sending Helix, as described............. **$4.00**
Shipping weight 7 lbs.

The "Electro" Adjustable High-Tension Condensers

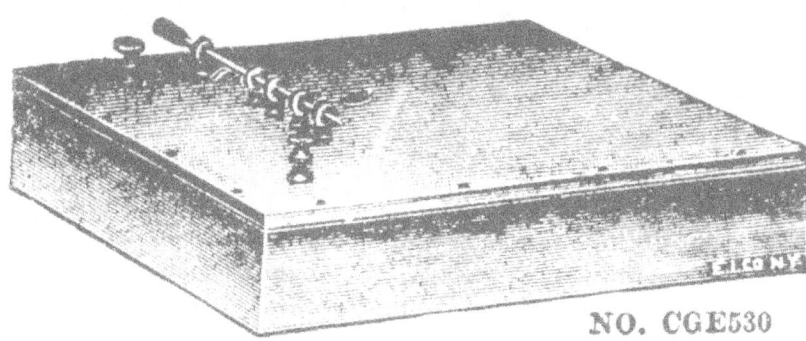

NO. CGE530

For many years we sold these fine condensers with no adjusting arrangement, but of late a heavy demand for a high tension adjustable condenser has sprung up and we are therefore more than pleased to present our condensers now with the adjustable feature.

The connections are made in such an ingenious manner that either one, two, three or all nine or nineteen plates can be put in circuit, simply by sliding the contact rod into more or less contact bushings. The adjustment is quick, sure and easy, no switches or levers need be touched. The adjustable feature is of incalculable importance for wireless work, as no spark coil, transformer coil, or transformer can work to the highest efficiency without the right capacity, which can only be obtained by means of a condenser with a variable capacity. No. CGE530 has 5 contact bushings, No. DGE531 has 10 contact bushings.

The construction is simple and durable, and sparking is absolutely prevented. The cases are solid quartered oak, highly finished. For dielectric we use imported French glass sheets of a special grade, 1-16 in. thick, free from salts and air bubbles. Instead of tinfoil we use metal plates. The No. CGE530 has 1440 sq. in. of active condenser surface. It can be used up to ½ K.W. Size over all is 11¾x14¾x2¾ in. Its maximum capacity is .009 microfarads.

The No. DGE531 has 19 metal plates, 3040 sq. in. of active condenser surface. It can be used up to 1 K.W. Sizes, over all, 11¾x14¾x3¾ inches. Its maximum capacity is .0203 microfarads. Both condensers are sealed in with a large amount of Pure Sealing Compound, which not only prevents bursting of the plates, but also safeguards the condenser from breakage, and, to a large extent, from puncturing. Two heavily nickeled binding posts are furnished. Each condenser is fully guaranteed as to capacity.

No. CGE530 High Tension Adjustable Condenser, as described.. **$3.75**
Shipping weight 40 lbs.
No. DGE531 High Tension Adjustable Condenser, as described.. **$4.75**
Shipping weight 45 lbs.

The "Electro" High Voltage Condensers

No. AX2345

Originally developed for use in our "Intercity" outfits. Each condenser consists of 4 glass plates, 4x4¾ in. in size, with metal sheets in between. These condensers are exactly right for use with a 1 in. spark coil but may be built into sections to accommodate most any source of high frequency current. Their convenient form makes them especially adaptable to most any apparatus. They are cheaper and occupy less room than Leyden jars. Our cut shows only the outside of the condenser, which we supply SEALED UP IN A SPECIAL COMPOUND, with a nice Hardwood Box, finely finished. This absolutely prevents leakage or breakage. Size 4⅞x6½x1 in.

No. AX2345 "Electro" High Voltage Condenser, as described.. **$1.00**
Shipping weight 1 lb.

The "Electro" Adjustable Sending Condenser

PATENTED MARCH 8, 1910

It has long been known that by connecting Leyden jars (a capacity) across the spark gap, the sending radius of a station could be increased enormously. The waves sent out from the regular spark gap die out very fast (Fig. 1), while the waves emitted from Leyden jars set up oscillations of much longer duration (Fig. 2), besides being more powerful. The latter is easily proven by connecting our Adjustable Condenser across the spark gap. Ordinarily the spark in the gap has a reddish-violet color, making little noise.

Now connect our condenser across the gap and you immediately notice an intense blue mass of fire. The noise even on a ½-inch coil is loud enough to be heard far away, while the noise from a 1-inch or 2-inch coil can be heard for blocks. It sounds what it means—"Business."

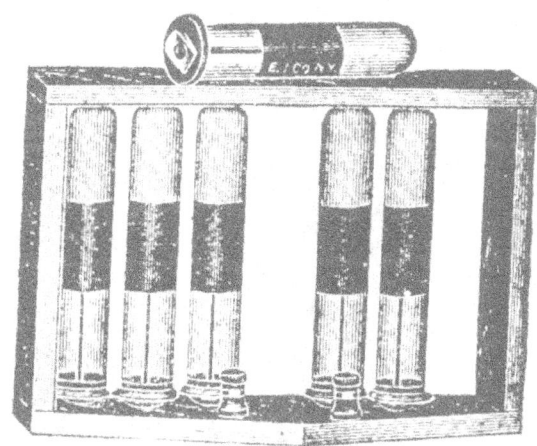

NO. BEK9260

The "Electro" Adjustable Condenser is a marvel of simplicity and efficiency. It is not alone used in wireless, but in Tesla experiments; in fact, in all high tension work, to adjust capacities, etc.

A complete Condenser comprises the stand or frame and six best imported Leyden jar condensers in which only glass free of all traces of lead is used. Leakage absolutely impossible. Each jar makes spring contact at the top, and as the stand has at the top circular recesses, and at the bottom small round metal indentations, the jar is snapped into its position in less than a second. It snaps out simply by pulling or pushing the jar.

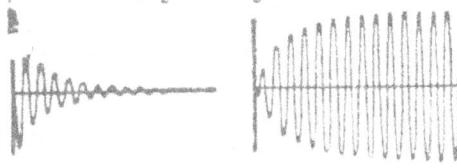

FIG. 1   FIG. 2

NO CONNECTING WIRES, NO SCREWS USED WITH JARS.

Good connections at ALL times. The jars cannot fall out, no matter in what position.

To change or vary the capacity of your circuit, simply snap in or out more or less jars, till best results are reached. The frame is made of well seasoned oak, the jars are beautifully finished and when connected (from 2 up) they are automatically placed in multiple. This, we found, is the best arrangement.

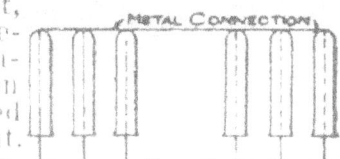

Sizes are: Height, 6 ins.; width, 2½ ins., length 9 ins.

No. BEK9260 "Electro" Adjustable Condenser, as described, complete with

Stand and 6 Leyden jars........................... **$2.50**
Shipping weight 2 lbs.

No. CE9262 Leyden Jar, each (Shipping weight 4 oz).......... **$0.35**

Dear Sirs:—   Lincoln, Nebr.

Last year I ordered a number of things of you, among which was an "Electrolytic" Interrupter for my wireless station. I have used it very hard all the while except during the summer. I found that it worked just as well after three months of non-use as it did when I got it. The "Telegraph Key" I received of you is a dandy, as is the Ball Bearing slider.   ASHLEY WILLIAMS.

The "Electro" Antenna Connector

NO MORE SOLDERED AERIALS. NO MORE LOOSE CONNECTIONS

No. CK3339

We present herewith the latest useful device to those erecting wireless aerials of any size; it is in the form of a brass terminal connector block for the lead-in or rat-tail juncture. The wiring diagram given indicates how the connector is employed to properly join the down-coming leads from the aerial to the lead-in wire, which may be No. 14 or larger. The larger the better. The rat-tail leads are made of No. 14 wire or cable as used in the aerial flat-top.

The weak point in most experimental radio plants lies in poor joints of the aerial, especially where it joins the lead-in wire. This weakness is avoided by using The "Electro" Antenna Connector, which assures perfect joints always.

This connector is made of solid and massive brass and solves the poor joint problem easily, both in an electrical and mechanical sense. The connector is provided with four No. 8/32 screws at the top for the rat-tails; proper size holes for the wires being provided. Also the rat-tails may be clamped under the 8/32 screw head if preferred. It is always best to solder or sweat in the leads, but the screws here provided make it possible to effect a perfect joint between the lead-wires without soldering.

The connector is drilled with a large and small hole at the base to accommodate any wire from No. 14 to No. 4 solid B. & S. conductor. A heavy No. 14-20 screw clamps this heavy wire. The connector, if desired, may be covered with friction tape after installing it, although this is not absolutely necessary. The size of the "Electro" Antenna Connector is 2 in. high by 1¾ in. wide by ⅝ in. thick.

No. CK3339 "Electro" Antenna Connector. Shipping weight 6 oz.   **$0.30**

Bamboo Spreaders

No. EK6527

These bamboo rods are 8 ft. long and taper slightly from 1¼ and 1½ in. at the butt. They are very strong and light. Must be sent by express unless cut in 2 ft. pieces.

No. EK6527 Bamboo Pole, 8 ft. long. Price each. Shipping weight 2 lbs.   **$0.50**

High Tension Cable

No. AE1298

It is used in most all wireless stations, and can stand an enormous high secondary discharge. The wires used in the construction of this cable are made from soft drawn copper, covered with three and four separate rubber insulations. It is very flexible, and can be handled easily, especially for laboratory and portable purposes. It is an absolute necessity to lead the antennae (aerial) from the station out in the open air.

The 5 millimetre (diameter) size will stand the discharge of our 1 in. coil. The 10 millimetre size will stand the discharge up to a 4 in. coil.

No. AK1297 High Tension Cable 5m/m Diam., per foot.......   **$0.10**
Shipping weight 1 oz. per foot.

No. AE1298 High Tension Cable 10m/m Diam., per foot........   **$0.15**
Shipping weight 2 oz. per foot. Not less than 5 feet sold.

WIRELESS CODES.

LETTERS	MORSE	CONTINENTAL	NAVY

ABBREVIATED NUMERALS USED BY CONTINENTAL OPERATORS.

This is a greatly reduced reproduction of our famous Wireless Code Chart, the original of which is 9x11 inches, printed on stiff cardboard that sells at $0.10. See page 77.

The Gernsback Electrolytic Interrupter

Patented April 4th, 1911

is a radical departure in electrolytic interrupter manufacture. It was constructed with the view to stand great abuse, gives marvelous results and to be ridiculously low in price. Heretofore such interrupters could not be had under $15 to $20 and most experimenters who did not care to pay this sum had to go on using batteries, which only cause trouble and dissatisfaction.

The Gernsback interrupter is connected in series, with any ordinary spark coil and the 110 V. direct or alternating lighting current supply. No resistance or condenser is used, except a key or switch to break the current in the usual manner. The vibrator of the coil must be screwed up tight as it should not vibrate. The glass vessel is filled with the solution (formula furnished only with interrupter), and as soon as the key is depressed you will get the surprise of your life. Instead of a thin, meagre spark, as with batteries, you get a HEAVY FLAME ¼ INCH THICK. That this is the ideal thing for Wireless is unnecessary to mention. The spark obtained of a 1-inch coil, connected to a big sending condenser and a zinc spark gap with zincs ½ INCH THICK will crash in the gap with such a tremendous noise that it will take your breath away AND THE SPARK FILLS THE GAP. These are PLAIN FACTS backed by our usual guarantee. By way of proving our statement look at the two photos taken by Mr. Gernsback. The first one shows the full spark of a 2-inch coil run by a 6 V. 60 A. H. storage battery. Exposure 1½ seconds. The second shows the FLAME of the same coil with a 110 V. current and the new interrupter. Exposure 1½ seconds. The flame shoots upward, as the great amount of heat raises the discharge. You are able to get a better and heavier spark

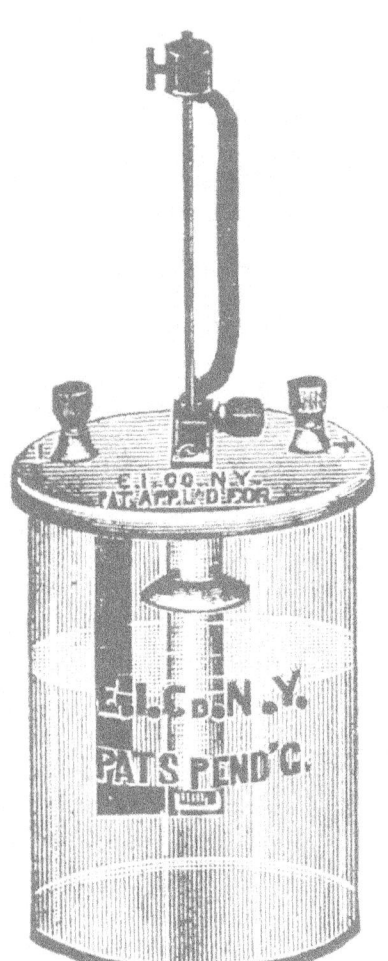

NO. BHE8000

from 15% to 25% LONGER, all depending on the construction of the coil.

And that is not all. The output of the coil is increased at least 60%. That means that you can send at least 60% further with the Gernsback interrupter. This will be better understood by mentioning that two No. 14 copper wires, connected to a 1 inch coil and separated ¼ inch will fuse within 5 to 10 seconds. The Gernsback interrupter starts at 50 volts. A metal rod of especial alloy goes through the cover down in the porcelain tube. This tube at its lower end has a peculiar aperture in which the pointed rod fits. The tube at the upper end has a threaded top which screws in the cover. This tube is made of special material and will not crack even if the interrupter is worked steadily. wears itself away to a point. The rod is fed down by gravity and is entirely controlled by the weight attached to the top of the rod (see ill.). In fact, the entire success of this interrupter lies in the right size of the metal weight. Too much weight gives no spark at all; too little gives an uneven and unsteady spark. New rods are supplied at a trifling cost. The rod can be left constantly in the solution.

NO. 1
ORDINARY SPARK.

In operation the metal rod

The porcelain cover has all metal parts IMBEDDED in it (patented). No metal is exposed whatsoever. Therefore NO CORROSION. The binding posts being of hard rubber cannot corrode, become short circuited accidentally, nor shock you. The interrupter heats up very little even with steady work. The path between the two electrodes is only ¼ inch and the amount of solution heated at a time therefore is necessarily very small.

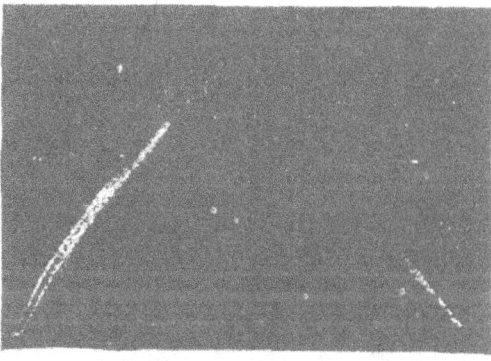

NO. 2
SPARK OBTAINED WITH INTERRUPTER.

This interrupter has found thousands of friends and is especially recommended for wireless and X-ray work. When used for wireless it may be stated that it produces an extremely high sound in the distant receiver, which is much easier to read than the low sound produced with the old spring vibrator giving only from 150 to 200 interruptions per second, against 5,000 to 7,000 per second with the electrolytic interrupter. The interrupter is to be used in connection with ordinary spark coils from ¼ inch up to 12 inch spark length, or our GGE8050 Transformer.

Two coils (or more) may be connected in series and if the secondaries are connected in series too, the length of the resulting spark is as long as the spark of the two coils put together. Therefore, two 2-inch coils will give a 4-inch spark and so on. Ordinary vibrator coils can not of course be connected in series, as each vibrator opposes the other, the spark length is cut down.

With the electrolytic interrupter a plurality of coils work as one, as the pulsations from the interrupter flowing through all the primaries (connected in series) magnetize and demagnetize the primaries all at the same time. The result, therefore, is that each coil acting in unison with the other (or others) will add its output to the other (or others). The longer spark is the result.

OPERATION

First fill the glass jar with the solution (to be obtained from any druggist) so that it stands 2½ inches from the top of jar. Put the cover on jar and pass the rod through the cover down in the tube. Be sure that its point fits in the aperture at the bottom of tube. The weight is then attached to the rod as shown in ill. The thumb screw of the metal bridge on top of cover is left loose. Now connect the interrupter as shown in diagram. If the current is direct the positive pole of the current must be connected with the post marked +. If the current is alternating it does not make any difference how the wires leading to interrupter are connected, since there is no positive nor negative pole.

The Interrupter works on direct and alternating current.

A switch block with fuses should always be used with the interrupter. It is much better to blow out a fuse than to damage the coil or interrupter if the current should get too strong, or if the tube in the interrupter should accidentally become fractured, which would short circuit the line.

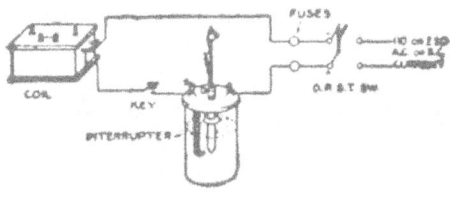

Every instrument is fully guaranteed to be all we claim for it. Mr. Gernsback would not allow his name put to it, if he had not implicit faith in it. It is a guarantee by itself.

Note.—This Interrupter does NOT work our closed core transformer, but only open core transformers, such as our No. GGE8050.

No. BHE8000   THE GERNSBACK INTERRUPTER, as described. **$2.85**
Size 10½x5 in. Shipping weight 6 lbs. Each......

No. BE8000a   Metal rods for Gernsback Interrupter. Each........ **$0.25**
Shipping weight 4 oz.

No. EK8000b   Interrupter Tube. Shipping weight 4 oz............ **$0.50**

No. DE8000c   Interrupter Jar, 4½x6½ in. Shipping weight 3 lbs.
Each **$0.45**

The "Electro" ½ K. W. Transformer-Coil

(100 MILE WIRELESS COIL)

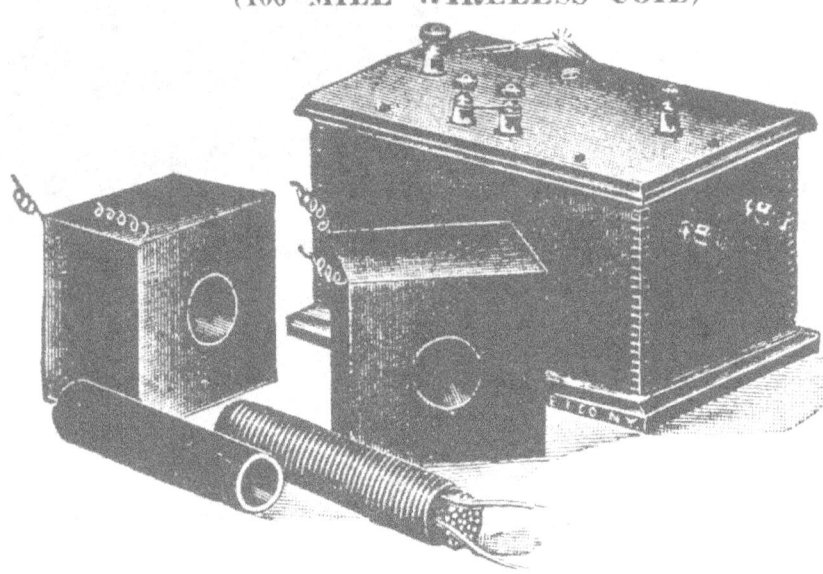

No. GGE8050

is a radical departure from ordinary coil building. It possesses all the good points of a good coil, but none of its bad ones.

The average experimenter when buying a coil nowadays buys a cat in a bag. The coil is sealed entirely and if it should break down it must go back to the factory. Neither does the owner know what is inside of the coil—he must take the maker's word for it. Our new coil is **NOT SEALED IN,** yet is better insulated than a sealed in coil. The new departure is our **BLOCK SECONDARIES** (see illustration). These secondaries are wound with **ENAMELED WIRE.** This means, on account of getting 3 times as many ampere turns into a given space, that our secondaries are 3 times as efficient as other ones, and that they take up one-third as much room. Size of secondary, 3½x2¾x3½ inches. You marvel that such a small coil could give such an enormous output. The enameled wire explains the mystery. After the secondary coils have been wound they are placed in a square mould which is filled with molten insulating compound. When cold, a square block-coil is obtained, which exposes no wire except the 2 connections. We form our secondaries square so they can not roll. Each secondary weighs 2¾ lbs. and gives a 1-inch spark. The primary is wound with Double Insulated Copper Wire No. 14, B. & S. and separated by a hard rubber insulating tube from the two block secondaries. The whole is placed in the coil box, which has been treated with an insulating compound. All coils fit snugly in the box and are arranged in such a way that they can not move and are always ¼ inch apart. After connections are made the cover is screwed down, and this marvel of simplicity is always ready to be inspected and to be taken apart, when occasion arises, for new experiments, etc., etc.

Four top metal binding posts are provided, so that one secondary may be used at a time, both in series, both in parallel and for other important experiments.

BY CONNECTING IN MULTIPLE, RANGE IS GREATLY INCREASED.

As there is no vibrator or condenser to this coil, it must, of course, be used with our electrolytic interrupter by running it from 110 Volts Alternating current, or 110 Volts Direct current.

The spark obtained is 1 to 1½ inches long, but ¼ inch THICK. For wireless work it is the fat spark that counts, not the long, thin spark. You must radiate (amperage) from your antenna, not tension (voltage).

Here is an UNSOLICITED testimonial:

Electro Importing Co.   New Orleans, La., Jan. 11, 1911.

Dear Sirs: It may be of interest to you to know, that I communicate with a friend in Baton Rouge every night with my One-half K.W. Transformer Coil, a distance of about 70 miles air line. My coil is working excellent, and anyone wishing to buy a coil, cannot make a better move than by purchasing one like mine. Hoping this will be good news, I remain,   Very truly yours,

BERNARD OPPENHEIM, 1435 Henry Clay Ave., New Orleans, La.

Continued on page 105

INDEX

A
Page
Aerial Construction 13
Aerial Insulators 81, 82
Aerial Wire 74
"Amateur" Wireless Phones 69
Ammeters 142, 143
Ammeter, Hot Wire 87
Antenna Connectors 92
Antenna Insulators 81, 82
Antenna Switch 83
Antenium Wire 74
"Arlington" Baby Timer 25
Assortment, Mineral 77
Automatic Charging Cut-out 151

B
Back-plates, Carbon 170
Ball, Zinc Spark 144
Bamboo Spreaders 92
Battery, Chromic 135
Battery Connectors 165
Batteries, Dry 144
Battery Meters 142
Battery Motors 155
Batteries, Storage ....136 to 141
Beakers 121
Bell Ringing Transformers 169
Belts, Leather 153
Blowpipe 124
Bottles, Glass 118, 123
Books 171, 172
Bornite 76
Bound Volume "Experimenter" 170
Boys' Electric Toys 133
"Boy Scout" Wireless Outfits.......24
"Bull Dog" Spark Coils......106 to 114
Bunsen Burner 159
Burette Clamp 125
Button, Microphone 160
Buzzers 61, 78, 79, 129
Buzzers, Radiotone 61

C
Cabinet Radio Outfits............25 to 33
Cable, High Tension................92
Carbon Balls 144
Carbon Diaphragms 170
Carbon Grains 144
Carbon Grain Transmitter 82
Carbon Rod 135
Carborundum 77
Cardboard Tubing 43
Cells, Dry 144
Cells, Selenium 158
Charging Cut-Out, Automatic 151
Chemical Laboratory 116
Chemistry 115
Chemicals 126
Chromic Plunge Battery 135
Clamp, Ground 60
Closed Core Transformer 105
Code, Wireless Chart 77, 93
Codophone 130
Coherer, Precision 79
Coils, "Bull-Dog" Spark......106 to 114
Coils, Loading 52, 53
Coils, Transformer 96
Coils, Tuning 46, 47
Compass, Magnetic 168
Condensers 55-58
Condensers, Adjustable Sending 91
Condensers, High Capacity 79
Condensers, High Tension 90

D
Page
Connectors, Antenna 92
Connectors, Battery 165
Connectors, Separable Wire 168
Contacts, Telegraph Key 86
Copper Pyrites 76
Cords, Telephone 74
Core, Induction Wire 157
Couplers ....48 to 51
Course, Wireless ....100 to 104
Crucibles 123
Crystals 38, 76, 77
Cut-Out Automatic 151

D
"DeLuxe" Radio Crystals 40
Desiccating Jar 119
Detectiphone 161, 162
Detectors, Wireless 41, 44, 45
Detector Crystals 38, 76, 77
Dials for Omnigraph 127, 128
Diaphragm, Receiver 74, 170
Direct Current Motors 155
Discharger, Leyden Jar 157
Double Pole Receiver 71
Dry Cells 144
Dynamos 152, 154, 155
Dynamo for Lighting Plants........152

E
Electric Buzzer 78, 79
Electric Motors 155
Electric Toy Outfit 133
Electrite Specialties 80
Electro-Magnet 144, 165
Electrolytic Interrupter 94
Electrolytic Rectifier 145
Electrodes, Vacuum 164
Electrose Insulators 81, 82
"Electro" Telegraph Outfit 131
Evaporating Dishes 120
Engine, Gasoline 152
Erlenmeyer Flasks 119
Experiment Spark Coils .....111 to 113
"Experimenter" Bound Volume.......170

F
Ferron 76
Filter Paper 123
Filter Pump 124
Filter Stand 126
Fixed Condensers 55, 57
Flasks 118, 119
Foil, Tin 78
Frequency, High, Outfits 163, 164
Friction Tape 157
Funnels 119, 121

G
Galena 76
Galena Detector 44, 45
Galvanometer 150
Gap, Zinc Spark 83, 84
Gasoline Engine 152
Glass Bottles 118, 123
Glassware ....118 to 123
Glass Spirit Lamp 159
Gold Leaf 160
"Government" Wireless Phones......66
Graduates 119, 120
Ground Clamp 60
Ground Wire 82

H

	Page
Handles	80
Hawkins Electric Guides	171, 172
Headbands, Phone	66 to 72
Helix, Sending	89
Hercules Dynamo	155
High Capacity Condensers	79
High Frequency Apparatus	163, 164
High Tension Cable	92
High Tension Condensers	90
Hot Wire Ammeter	87
House Lighting Plants	150 to 153
"Hugonium"	45
Hydrometers	141, 142

I

Induction Coils	106 to 114
Induction Core Wire	157
Instruments, Laboratory	118 to 125
Insulators, Antenna	81, 82
Insulators, Porcelain	81
"Intercity" Radio Outfit	24
"Inter-Ocean" Radio Outfit	21
Interrupter, Electrolytic	94
"Interstate" Receiving Outfit	22
Iron Pyrites	76

J

Jars, Leyden	87
Jars, Specimen	118
"Junior" Fixed Condenser	55
"Junior" Tuner	47
"Junior" Wireless Phones	70

K

Keys, Telegraph	85, 86, 131
"Key-West" Radio Outfit	26
Keys, Wireless	86
Kickback Preventer	86
Knobs	80

L

Laboratory, Chemical	116
Lamp, Glass Spirit	159
Lead, Peroxide of	77
Learners' Practice Set	129
Learners' Telegraph Outfit	131
Leyden Jars	87
Leyden Jar Discharger	157
Lighting Plants	150 to 153
Lightning Switches	82
Loading Coils	52, 53
Loose Couplers	48 to 51
Loud-Talker	162
Low Voltage Transformer	169

M

Magnets, Electro	144, 165
Magnetic Compass	168
Meters, Battery	142, 143
Meters, Hot Wire	87
Meter, Wave	59
Microphone Button	160
Microphone	82
Minerals, Wireless	38, 76, 77
Molybdenite	77
Mortars	121
Motors, Electric	155
Motors, Low Voltage	155
Motor Type "SS"	155

N

	Page
"Nauen" Radio Outfit	32
"Navy Type" Loose Coupler	50
Nickel Plating Outfit	149

O

Omnigraphs	127-128
Order Blank	175, 176
Oscillation Transformer	88
Outfit, Nickel Plating	149
Outfits, Sending	24 to 37
Outfits, Wireless Receiving	21 to 37
Outfit, Soldering	168

P

Panel Radio Outfits	25 to 33
Paper, Filter	123
Parcel Post Rates	5, 6
Perikon Mineral Set	77
Peroxide of Lead Tablets	77
Phones, Wireless	66 to 72
Pinch-Cock	125
Plants, House Lighting	150 to 153
Plating, Nickel	149
Pocket Meters	142
Pony Receiver	73
Porcelain Dishes	120
Potentiometer, Rotary	75
Practice Set	129
Precision Coherer	79
Preventer, Kickback	86
Professional Loose Coupler	49, 50
Pyrites, Copper and Iron	76

R

Radiocite	38
Radiocite Detector	41
Radioson Detector	39
Radiotone Buzzer	61
Radio League	19
Radio Outfits	21 to 37
Raw Material	166, 167, 168
Reagents	126
Reagent Bottles	123
Receiver Cords	74
Receiver Diaphragm	74, 170
Receiving Outfits, Wireless	21 to 72
Receivers, Wireless	66 to 72
Rectifier, Electrolytic	145
Reducer, Current	146
Regulator, Rheostat	146
Rheostat	146
Rhumkorf Coils	106 to 114
Rotary Potentiometer	75
Rotary Spark Gap	84
Rotary Var. Condenser	56, 57, 58, 59
Rubber Tubing	122

S

Sand Bath	124
Salt, Nickel Plating	149
Selenium Cells	159
Selenium Metal	159
Sending Condensers	90, 91
Sending Helix	89
Sending Outfits	24 to 37
Silicon	76
Sliders	54
Soft Metal "Hugonium"	45
Solderall	78
Soldering Outfits	168
Solenoid	185

	Page
Spark Coil, "Bull-Dog"	106 to 114
Spark Gap Balls	144
Spark Gap, Zinc	83, 84
Specimen Jars	118
Spreaders, Bamboo	92
Spirit Glass Lamp	159
Storage Batteries	136 to 141
Students' Chromic Battery	135
Switch Antenna	83
Switchboards	151
Switchboard Meters	143
Switches, Lightning	82
Switchhandle	80

T

	Page
Tape, Insulating Friction	157
Telegraph Keys	85, 86, 131
Telegraph Knob	80
Telegraph Machines	129, 130, 131, 132
Telephone Cords	74
Telephone Receivers	66 to 72
"Tesla" Transformer	147 to 149
Test Tubes Brush	125
Test Tubes	121
Test Tubes Holder	125
Test Tubes Rack	125
Therapeutic Apparatus	163, 164
Thermometer	123
Thistle Tubes	121
Thumb Screw	80
Time by Wireless	20
Tin Foil	78
Toys, Electric	133
Toy Transformer	169
"Transatlantic" Receivers	68
Transformer, "Bell Ringing"	169
Transformer, Closed Core	105
Transformer Coil	96
Transformer, Oscillation	88
Transformer, "Tesla"	147 to 149
Transmitter, Carbon Grain	82
Transmitter Button	160
"Trans-Oceanic" Loading Coil	53
Treatise on Wireless	8 to 19
Triangle, Clay	125

	Page
Tripods	124
Tubes, Cardboard	43
Tubes, Glass	122
Tubes, Rubber	122
Tuners	46, 47
Tuning Coils	46, 47
Tuning Sliders	54

U

	Page
Universal Detector Stand	45

V

	Page
Vacuum Electrodes	164
Variable Condensers	56 to 59
Vario-Selective Coupler	51
Violet Ray Machines	163, 164
Voltammeters	142
Volt Meters	142, 143

W

	Page
Wash Bottle	120
Watch Case Buzzer	79
Watch Glasses	119
Water Bath	124
Wave Meter	59
Wires, Aerial	74
Wire Gauze	124
Wires, Ground	82
Wire, Induction Core	157
Wireless Code Chart	77, 93
Wireless Course	100 to 104
Wireless Detectors	41, 44, 45
Wireless Key	86
Wireless Outfits	21 to 72
Wireless Receivers	66 to 72

Z

	Page
Zincite	76
Zinc Spark Ball	144
Zinc Spark Gap	83, 84
Zinc Rod	135

FREE — Wireless Correspondence Course
IN 20 LESSONS
By
S. GERNSBACK, A. LESCARBOURA and H. W. SECOR, E. & R. Eng.
FREE

In Use on Every Battleship of the U. S. Navy

Let Us Help You to Become a Practical WIRELESS EXPERT

THIS IS THE ELECTRICAL AND WIRELESS AGE

Wonderful opportunities are offered to the man who has special training to-day. We will help you to become an expert in Wireless. Hundreds have done it, why not you? Just try it! You can do it as well as anybody else!

Our aim is not only to sell you electrical goods! We want to instruct you how to handle to the best advantage all the Wireless Apparatus; to tell you the how and why of the fascinating art of Wireless. This is the Reason for offering you our

FREE COURSE.

Of course, we want you for a customer. You are interested in buying electrical and wireless apparatus. Why not be one of our regular patrons? We promise to give you the maximum of quality, plus service and to show you OUR appreciation, we are going to give you a profit sharing PREMIUM on every dollar you spend with us.

Read the following proposition. It means an absolutely unique opportunity.

CONDITIONS:

In the following pages you will find twenty Wireless Course Certificates.

When sending your order which must not be less than One Dollar ($1.00), attach one of these coupons and you will receive with your goods the first lesson of the WIRELESS COURSE and a superb cloth binder.

With every following order amounting to not less than $1.00, we send you another lesson. This means that for every dollar's worth of goods you order, you will receive as a premium, one lesson of our

"WIRELESS COURSE"

It is understood that you can receive the whole course with

The Beautiful Cloth Cover of our Wireless Course.
Size 6½ x 9¾ inches.

one order if same amounts to $20.00, or ten lessons, if you order goods amounting to $10.00, etc.

You may, of course, select any number of the Wireless Course you want; but the cloth binder is only furnished with lesson Number One.

☞ ABSOLUTELY NO DISCOUNT ALLOWED FROM ANY ORDER APPLYING TO THIS FREE WIRELESS COURSE, AND NO LESSONS WILL BE SENT ON ANY ORDER UNLESS THE CORRESPONDING COUPON IS ATTACHED.

It is needless to say that our Wireless Course is up-to-date and absolutely thorough, commencing with the most complete

Contents: 160 pages, 350 illustrations, 30 tables.

explanations on Electricity in General and Wireless, ending with chapters on scientific mathematics and complete history of Wireless.

Do not miss this splendid opportunity.

Send in your order and your Wireless Course Certificate TO-DAY.

READ contents of Wireless Course on back of coupons.

This course cannot be bought and can only be obtained under the above conditions.

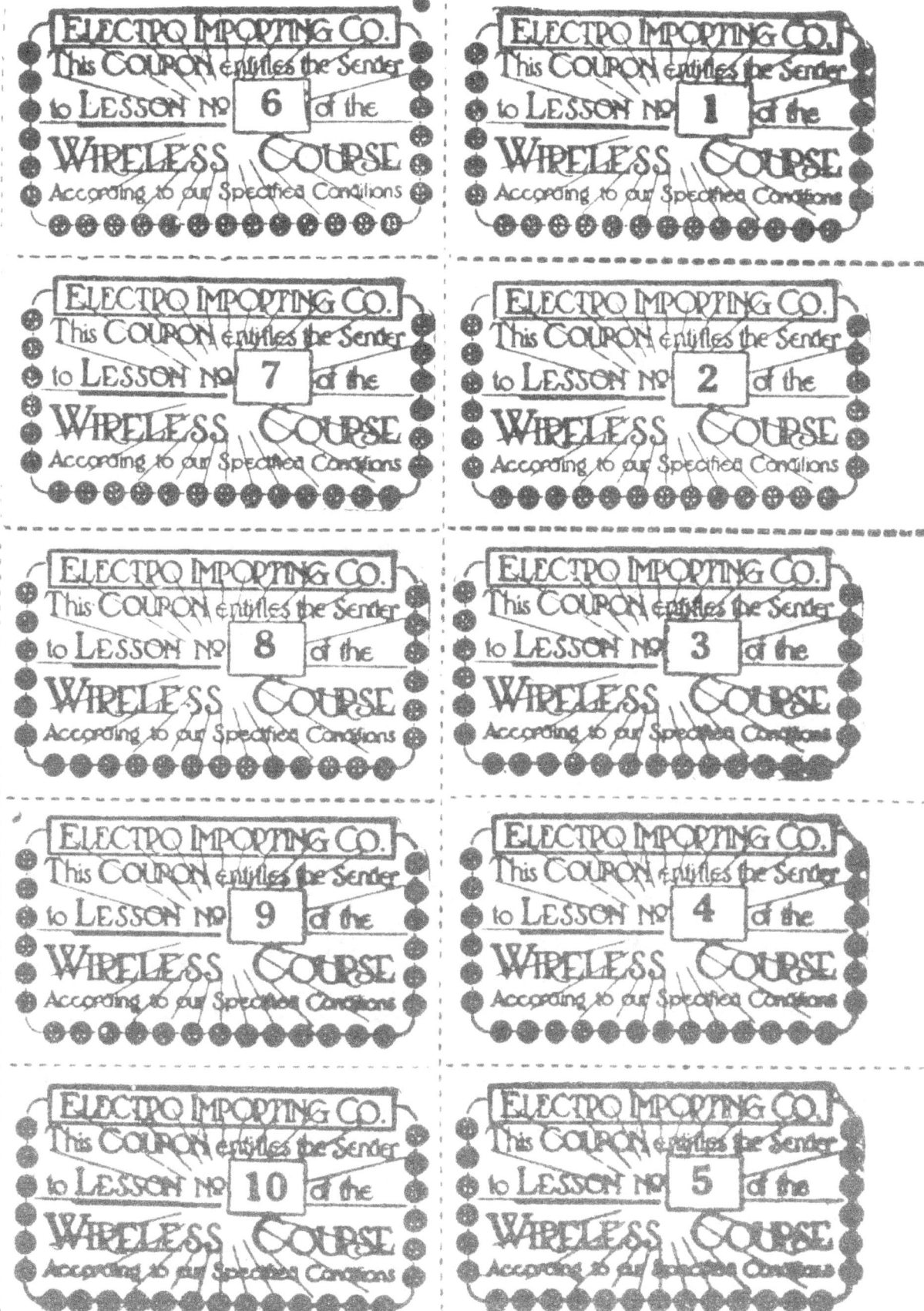

102

Contents of Lesson No. 1

THE PRINCIPLES OF ELECTRICITY.

Conductors, Insulators, Static Electricity, Current Electricity, Electro-Motive force, Batteries, Measuring Instruments.

Contents of Lesson No. 2

THE PRINCIPLES OF MAGNETISM.

Compass, magnetic flux, Electrodynamics, Electro-Magnetic Induction, Induction Coils and Transformers.

Contents of Lesson No. 3

DYNAMOS, MOTORS, GENERATORS AND WIRING

Series Motors, shunt motors, compound motors, Power transmission and wiring. Transformer boxes. Frequency, Edison 3-wire system.

Contents of Lesson No. 4

THE PRINCIPLES OF WIRELESS TELEGRAPHY.

Maxwell's theory, Hertzian waves, Branly's coherer. G. Marconi. Principles of wave length, tuning, oscillating circuit, autotransformers, etc.

Contents of Lesson No. 5

THE AMATEUR TRANSMITTING SETS AND APPARATA.

Spark Gaps, Coils, Sending Helices, Condensers, Interrupters, Keys.

Contents of Lesson No. 6

TRANSMITTING SETS (Continued)

Aerial switch, commercial stations, Motor-Generator, Heavy keys, commercial spark-gaps, Rotary spark-gaps, Leyden Jar Condenser, Lightning switch, commercial wireless stations.

Contents of Lesson No. 7

NEW TRANSMITTING SYSTEMS.

Quenched Spark system, Telefunken system, Poulsen system, Duddel Arc, Complete Poulsen Station.

Contents of Lesson No. 8

RECEIVING APPARATA.

Detectors, Tuning of Receiving Apparata, Loose Couplers, Variometers, Detectorium, Portable Receiving Set.

Contents of Lesson No. 9

RECEIVING APPARATA.

Variable Condensers, Rotary Condensers, Fixed Condensers, Potentiometers, Wireless Receivers, commercial receiving sets.

Contents of Lesson No. 10

THE DETECTORS.

Coherers, Slaby-Arco-Vacuum coherer, Branly Detector, Automatic Detectors, Relays, Crystal Rectifiers, Silicon—Perikon—Galena—Molybdenite—etc., detectors, Electrolytic Detectors, Peroxide of Lead Detectors, Fleming Valve, Audion.

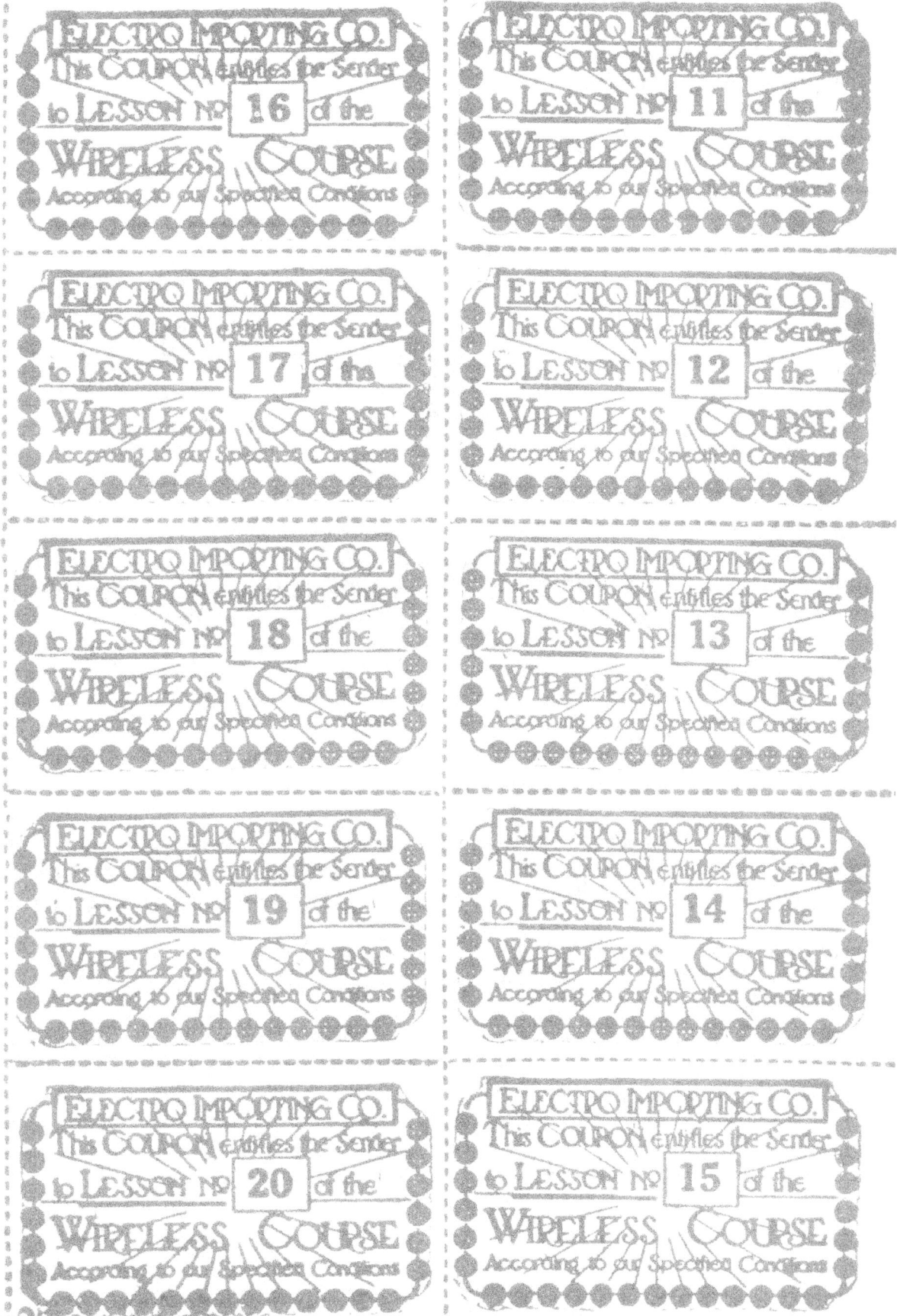

Contents of Lesson No. 11
AERIALS.

Antenna, Wiring, Insulators, Looped Aerial, Umbrella aerial, Lead-in, Bellini-Tosi Radiogoniometer, construction of aerials, etc.

Contents of Lesson No. 12
THE HOOK-UPS AND CONNECTIONS.

Study of the diagrams, Wireless telegraph symbols, close coupled systems, connecting interrupters, shipboard stations, Fessenden station, Receiving sets, Loose-coupled sets, Marconi selective receiving set, etc.

Contents of Lesson No. 13
THE HOOK-UPS & CONN. USEFUL INFORMATION.

Fessenden interference Preventer; Telefunken receiving set. Duplex Receiving set. The Collin system. The Lee De Forest system, Dielectric strengths of insulators, Notes on Ropes, Equivalents, connecting and soldering wires, Electrical units.

Contents of Lesson No. 14
OPERATION OF THE INSTRUMENTS.

Wave-length, Wave-Meters, Tuning. The use of the different instruments, Wireless Regulation.

Contents of Lesson No. 15
LEARNING TO OPERATE. THE CODES.

Operating the key, patent keys. The codes, Omnigraphs. The different codes, cipher messages, Abbreviations, Government messages, commercial messages. The Wireless Law.

Contents of Lesson No. 16
COMMERCIAL SHIP AND LAND WIRELESS STATIONS.

The Nauen station, United Wireless station. War ship stations. Commercial ship stations. Army stations.

Contents of Lesson No. 17
HIGH FREQUENCY CURRENTS.

Tesla experiments. Prof. Fessenden's experiments. Tesla Transformer, Oudin Transformer.

Contents of Lesson No. 18
THE WIRELESS TELEPHONE.

The principles. Collins system. Poulsen's system, etc.

Contents of Lesson No. 19
THE MATHEMATICS OF WIRELESS TELEGRAPHY.

Calculation of wave-lengths, Inductive calculation, Capacity calculation, Range of stations, Tables, Data, etc.

Contents of Lesson No. 20
THE HISTORY OF THE DEVELOPMENT OF WIRELESS TELEGRAPHY.

Steinheil, Edison, Maxwell, Tesla, Hertz, Crookes, Hughes, Popoff, Marconi, etc., etc.

Our coil radiates energy—high amperage—and lots of it. Compared with the ordinary coil, ours, as far as wireless transmission is concerned, will send further than the 8-inch coil wound with No. 36 B. & S. wire. And an 8-inch coil costs $95.00. Size of box, 9x5½x4½ inches.
No. GGE8050 Electro ½ K.W. Transformer-Coil, as described.... **$7.75**
Shipping weight 12 lbs.

The "Electro" Closed Core Transformer

TRIPLE CONTROL TYPE

Commercial and experimental wireless stations to-day are using invariably a closed core alternating current transformer for sending messages. They are thoroughly reliable and of the highest electrical efficiency. Superior to spark coils, they are, besides, very cheap to operate; the cost of running the ¼ Kilowatt (KW.) size, being 2 cents per hour. They will not heat up even if run continuously.

These improved type "Electro" quality transformers operate on the same principle (induction) as spark coils, but the transformer is simply connected to 110-125 volt A.C. 60 cycle circuits, with a telegraph key such as our No. ABE1117 or BX9212 in series, to make and break the primary circuit.

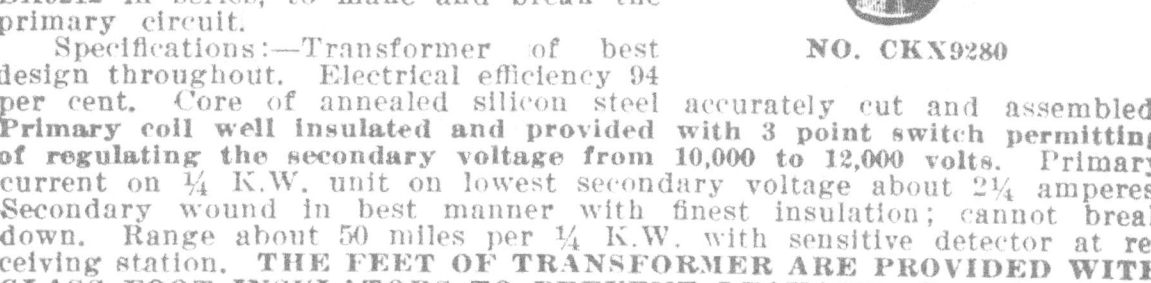

NO. CKX9280

Specifications:—Transformer of best design throughout. Electrical efficiency 94 per cent. Core of annealed silicon steel accurately cut and assembled. **Primary coil well insulated and provided with 3 point switch permitting of regulating the secondary voltage from 10,000 to 12,000 volts.** Primary current on ¼ K.W. unit on lowest secondary voltage about 2¼ amperes. Secondary wound in best manner with finest insulation; cannot break down. Range about 50 miles per ¼ K.W. with sensitive detector at receiving station. **THE FEET OF TRANSFORMER ARE PROVIDED WITH GLASS FOOT INSULATORS TO PREVENT LEAKAGE.** Secondary leads brought out through heavy pocelain insulators as shown.

This transformer is extremely efficient, reliable, and flexible in control. For 120 cycles frequency cost is the same as for 60 cycles, below. For lower than 50 cycles frequency, add 20 per cent. to cost here given.

No. CKX9280   "Electro" ¼ Kilowatt Transformer, as described size 6¾ in. high by 8 in. long by 8¼ in. wide......**$30.00**
Shipping weight 32 lbs.

No. DEX9281   "Electro" ½ Kilowatt Transformer...............**$45.00**
Shipping weight 45 lbs.

No. FKX9282   "Electro" 1 Kilowatt Transformer.................**$60.00**
Shipping weight 65 lbs.

Gentlemen:—   Buffalo, N. Y.

I am VERY MUCH PLEASED with your ½ K.W. Coil No. 8050. I have SENT 48 MILES UP THE LAKE TO THE CITY OF ERIE a boat equipped with the Clark Wireless system. The operator said I CAME IN AS LOUD AS C.H., THE PORT HURON STATION. (10 K.W.)
Yours truly,   H. SCHOEFFLIN.

The How and Why of "Bull Dog" Spark Coils Before Buying Any Coil, Read This:

To the average experimenter most spark coils look alike. As long as the coil gives a spark, he is satisfied, and, as a rule, when buying a coil he is guided mainly by price. The firm slicing prices most, gets the order. Quality seems to be the last thing the average experimenter or radio amateur thinks of.

Therefore, a few words on the subject might not be amiss. Speaking generally, there are two kinds of spark coils in America to-day; the box coil and the "Bull-Dog" coil. When we started the original experimenter's supply house, in 1904, we did not make our own coils, and for several years we sold the box coil, which is nothing more nor less than an ordinary gasoline ignition coil, made and built for ignition work, and nothing else. It has no style to speak of, is crude and is put together as cheaply as possible. The platinum contacts are very small and wear out quickly and all the parts are, as a rule, light. But the main part— the secondary, does not conform to modern high-grade coil practice. Nine times out of ten it is bare wire wound, with a minute air space between adjacent wire turns. The various wire layers are insulated from each other by a single wrapping of thin paraffine paper. This construction reduces the cost of the coil a great deal, at the expense of efficient insulation. Internal sparking, if the coil is strained, is rather the rule in such secondaries. They do not stand up for continuous work and total break-downs are common. That means practically a new coil, for the makers charge almost as much for repairing a burnt out coil as for a new one.

The most important part of the secondary, though, is the **size of the wire.** An enormous amount of copper wire can be saved by reducing the wire size and this is unfortunately the rule with some makers. The finer the wire, the less there is required for a given spark length. **Also the finer the wire, the thinner the spark.** It is obvious that for average experimental work, especially wireless work, such a coil is totally unsuited. For radio work it is not the spark length that counts, it is the THICK SPARK. Thus, a 6-inch thin spark will not transmit over 30 miles, while a ½ K.W. transformer giving but a 1-inch spark will easily transmit 100 miles and more. A long, thin, stringy spark is best compared to a fine thin stream of water. It has no power. A fat, heavy spark is like a thick stream of water. Its power is very great.

We could say a great many more things about the box coil, but we will refrain, because we might be thought of as "knockers." Suffice it to say that we discontinued the sale of the box coil because we lost a great many customers due to well-nigh universal dissatisfaction.

Mind you, we do not say that box coils are not good. There are excellent coils of that kind on the market—but the majority of them palmed off by certain houses are just plain ignition coils—very good coils for such work, but **never** for extended experimental or radio work.

You don't buy dry cells to light your house—you use a dynamo or storage cells. Each has its function. So with the box coil and the "BULL-DOG" coil. Each has its sphere of utility.

When we finally did start in the manufacture of the "BULL-DOG" coils we knew exactly what to avoid, in order to turn out a satisfactory article.

The many years of popularity and the wonderful sale that our "BULL-DOG" coils have enjoyed especially by high schools and universities, leads us to believe that we have the ideal spark coil for experimental and radio work. Naturally, our coils cost more than box coils—for we put more material into them. Thus our 1-inch "BULL-DOG" coil in OUTPUT equals most 1½ inch and 2 inch box coils. **Do not be misled by the length of a spark.** It means nothing; it is simply a trade trick. You do not buy a Mazda lamp bulb by its length; you want to know, primarily, **what its candle power is.** You pay accordingly. It is exactly the same with spark coils, the spark length is the second consideration. You want your coil to give a prodigious amount of power—not meaningless sparks. And if you use it for Wireless Work—and you will, sooner or later—you want to be sure that you radiate lots of power from your aerial. And only a genuine "BULL-DOG" coil does this.

COUNTERFEIT "BULL-DOG" COILS

There is only one "BULL-DOG" spark coil,—the "ELECTRO." **If it does not bear the "ELECTRO" name-plate, it is not a "BULL-DOG" coil.**

Certain unscrupulous supply houses, envious of our enormous success of these coils, have of late been marketing a coil which they palm off as the "BULL-DOG" type of coil. In appearance it looks like our coil, but it is nothing but an ordinary thin spark ignition coil, resembling externally somewhat our "BULL-DOG" coil. It does not meet our claims and is therefore spurious. For that reason also it sells at a lower price. But it is not a genuine "BULL-DOG" coil. It is to be deplored that competing houses, deliberately try to mislead the public with such antiquated business practices.

CONSTRUCTION

Only the very best material that money can buy is used in "BULL-DOG" coils. Although thousands of these coils are sold each year, not more than 4 have ever been returned to us in the lapse of any one year, due to broken down secondaries. And these cases have invariably been traced to straining the coil unduly either by using too heavy a battery current or by working the coil without having a load on the secondary.

SECONDARY

The heart of the coil. We use only enamel copper wire—nothing else—in our coils. This cuts down all leakage between adjacent wire turns. Individual layers are insulated with a heavy impregnated paper of highest insulation value.

No. 38 or 39 B. & S. wire is commonly used in most box coils. We use nothing thinner than No. 34 B. & S. electrolytic copper wire. **This means that "BULL-DOG" coils have from 25% to 50% more weight in wire than ignition box coils.** This costs a great deal more, but makes for powerful sparks.

All "BULL-DOG" Spark Coil secondaries, after winding, are boiled for a long period in a paraffine wax bath **under vacuum** to dispel all air from the windings. A costly process—but it keeps coils from coming back to us.

PRIMARY

All primaries are made of double cotton covered copper wire of suitable size. They are wound on impregnated insulating tubes and after winding are boiled in paraffine to expel all air and moisture. All our primaries are lathe wound and must be wound perfectly even to pass inspection. The core wire is imported from Norway. Only genuine Norway double annealed core wire is used by us. It costs 25% more, but the spark obtained is vastly better than if the domestic wire were used.

CONDENSER

We consider the condenser one of the most important parts of the "BULL-DOG" coil. If the condenser is not just right, if it is not **exactly** balanced, there will be excessive sparking at the vibrator contacts. Too large a condenser, while cutting down the vibrator sparking, also cuts down the secondary spark length. Too small a condenser increases vibrator sparking and decreases secondary spark length. Furthermore, as the condenser is subject to extraordinary stresses, due to the "back-kick" of the primary, its insulation can never be too good. Most cheap coils have poor condensers, often breaking down even if the coil is not abused.

"BULL-DOG" coils practically never have condenser trouble. In six years we did not replace more than eight condensers due to puncturing—a really wonderful record.

In the construction of our condensers, we use only the finest obtainable tin foil,—not lead foil—thick enough so it can't tear when the condenser is compressed hydraulically, after having been boiled in paraffine for some hours.

The paper we use is the best imported homogeneous rice paper, treated with paraffine **under vacuum**, to expel all air from its pores. We do not use wire leads, but broad copper strips to make connection with the tinfoil. This insures a perfect contact, and no resistance at the contacting surfaces. After treating in a paraffine bath, the condenser is compressed in a powerful press and a strong, perfect unit, almost impossible to puncture by the "back-kick" of the coil is the result.

After the primary has been provided with its insulating tube, the secondaries are slipped over this tube and the whole unit is placed in the outer enclosing tube. The highest grade of insulating compound is now poured around the unit, thoroughly sealing it in the enclosing tube. Over the enclosing a rich looking, fibre tube is slipped, and the coil is ready for assembling, after it has undergone several tests for spark length, break-down resistance, etc., etc.

Our small coils, such as our ¼ inch, ½ inch and 1 inch types, have the condenser enclosed in the outer enclosing tube. This condenser is of course insulated with great care from the secondary by at least 15 layers of Empire paper.

Our larger "BULL-DOG" coils beginning with the 1½ inch type have a sub-base in which the condenser is housed; it is surrounded by insulating compound.

MAHOGANY FRAME

Beginning with our 1917 model all of our "BULL-DOG" coils are encased in beautiful mahogany frames, piano-finish, hand rubbed. A more beautiful appearing coil can not be imagined—it shines and sparkles and is easily the most imposing piece of apparatus on your instrument table. You feel proud to own such a coil; its like is not found anywhere the world over—a pretty strong claim, but true, once you see a real "BULL-DOG."

VIBRATOR

We could write several pages about the superiority of our wonderful French style vibrator. It is practically noiseless and works at an astonishing speed. No large iron head retards its speed and gives a stringy spark. Our vibrator, on account of its great speed, gives a fat powerful FLAME-LIKE DISCHARGE at the secondary terminals, unlike anything you have ever seen. The vibrator is composed of one thick and one very thin spring and works exactly like a reed, the lower end of which has been clamped in a vise. Our vibrator when working emits a high NOTE, not a NOISE, as does the ordinary Rhumkorff or box coil vibrators. IT ALSO USES LESS CURRENT THAN ANY OTHER COIL.

TENSION ADJUSTMENT:

The "Bull-Dog" coil is the only one having a spring tension adjustment. By tightening screw 1 (see illustration) and by slightly loosening screw 2, the coil uses a minimum of current. By reversing this operation the coil draws its maximum current and the longest and most powerful spark is obtained. With the former adjustment the vibrator works fastest, with the latter adjustment it works slowest. Beautiful regulation is thus accomplished, not found in any other coil.

ADJUSTING THUMB SCREW:

Of course, while adjusting, the thumb screw must be regulated, till the secondary spark is satisfactory. Never screw the thumb screw too tight, as otherwise the contacts are apt to spark violently, thereby reducing their life. The coil works best when the sparking at the contacts is not too strong—just a mild, nice, blue spark, which should not blind you. One can tell at once if the battery used on the coil is too strong by observing the vibrator spark. If it sparks too much the battery current must be cut down at once, otherwise the secondary might burn out.

Our vibrator thumb-screw has several unique features. Once adjusted the vibration of the spring can never cause it to loosen, because we use a stiff, coiled phosphor spring under the head of the thumb screw. With the "BULL-DOG" coil you can not get a dangerous shock while adjusting the thumb screw, because THE RIM OF THE SCREW IS THOROUGHLY INSULATED with a thick, knurled fibre ring. If you have ever tried to adjust an ordinary spark, the secondary of which was grounded, (as in Wireless Work), you will appreciate this improvement, not found on any other coil.

CONTACTS

We now use only pure TUNGSTEN contacts in all our vibrators. Tungsten is the most refractive metal known to science to-day. It is very much harder than platinum and lasts at least 125% longer than the latter. The sparking on tungsten points is less than on platinum points and we very seldom hear of such a thing as a burnt out tungsten point. Inasmuch as the smallest tungsten point we use measures 1/8 inch in diameter and .05 inch thick (platinum points seldom are more than .100 inch in diameter and .0312 inch thick) the sparking is cut down to a minimum. Of course large tungsten points as these are more expensive than small platinum points, but we obtain better results, besides, large points last very much longer.

Due to the unique construction of our vibrator and due to the fact that tungsten metal does not melt readily, OUR VIBRATOR CONTACTS DO NOT "STICK" UNDER ANY CIRCUMSTANCES.

TESTING

All "BULL-DOG" coils undergo an elaborate series of tests before they are stamped with our inspection stamp. All our coils are tested twice; once before they are assembled, and once after assembling. The spark length must be right. It must be above the claim we make for that particular coil. Thus our 1-inch coils usually give 1¼ to 1½ inch spark. All coils are "strained" purposely to ascertain their break down resistance. We thus make sure that when you strain your coil accidentally it will not break down. Those that do break down at the factory once in a while are discarded. Only perfect coils leave our factory.

GUARANTEE

Read it carefully. It is the only coil guarantee made by any manufacturer, anywhere in the world to-day. It is the first time in the history of coil making that any firm ever attempted such an unheard of guarantee. It proves our supreme faith in our coils.

OUR GUARANTEE

We will replace without question any "BULL-DOG" coil within one year after its sale, providing it shows no gross abuse. In this guarantee is included partial or total burn out of the secondaries or broken down condenser, or any other mechanical defect arising, due to imperfect workmanship. This guarantee does not cover burned out tungsten vibrator contacts, which itself is a proof of gross abuse. Defective coils can only be replaced free of charge, if returned prepaid to our factory.

Now, of course, all this sounds nice and reads well. If you are still unconvinced read the following. Then if you want the best coil in America, we know that we will get your order:

Dear Sirs:— Alameda, Cal.

I tested the one-inch coil with a 6 volt, 6 ampere dynamo, and could with point dischargers, OBTAIN A STEADY SPARK OVER ONE INCH LONG. COULD SEND OVER SIX MILES, which took me by surprise, the aerial was 32 feet long, made up of four copper wires, on spreaders two feet wide, suspended from a mast 80 feet high. The detector WORKED FINE; COULD PICK UP SIGNALS without any trouble.

Yours truly, F. ARNBERGER, JR.

Gentlemen:— Chicago, Ill.

I bought one of your ½-inch spark coils and think it is a dandy. With four batteries I can obtain over ½-inch spark. CLARENCE MUELLER.

When ordering one of our Spark Coils or one of our Transformers, let us send you free with our compliments, lesson No. 2 "The Principles of Magnetism" of our famous "WIRELESS COURSE," giving you all the instruction about this apparatus.

Just attach coupon No. 2 to your order. For further information, consult colored section of this catalog.

The "Electro" "Bull-Dog" Spark Coils

It is a well known fact that we are to-day supreme in spark coil manufacture. We can prove that we sell more spark coils for experimental work than any two other concerns in the world.

The new "Bull Dog" type is the outcome of our 12 years' experience in this work, and for workmanship and appearance it stands unmatched.

We departed from the old, cold looking box style and now enclose

INSULATED THUMB SCREW

1
2

2-inch SPARK

$9.60

NO. IFX1089

the primary secondaries and condenser all in a fibre tube, enhancing the appearance a great deal and also making the coil far more compact and lighter at the same time.

A new French double spring vibrator with **double adjustment** is used now, giving extremely fast vibrations. The insulation is superb, internal sparking is impossible, as the greatest care is exercised to insulate all parts with the most expensive sealing compound. Our coils are especially constructed for use in wireless telegraphy, and we have devoted considerable labor and time in experimenting to produce something that we can recommend confidently to our customers. Our aim has been to furnish a coil not easily injured, even by rough handling, and these coils may be subjected to considerable rough usage without injury. The usual form of Ruhmkorff coil we found was too delicate and easily put out of order, and we therefore do not manufacture same. All the good features of the Ruhmkorff are embodied in our coils, and we guarantee them to give a very powerful and "fat" spark impossible to obtain with any other coil. All the experiments cited on the following pages can be performed with our coils, and we guarantee our coils to give better results for a longer period than any other coil.

We employ a condenser of large capacity in parallel with the vibrator, which decreases the sparking thereof as much as possible. All our coils have condensers, even the ¼ in. one. The vibrator contact points are of tungsten and will last almost indefinitely, providing the coil is not abused by the use of too strong a current.

If we say a coil gives 1 in. spark, this means that a 1 in. spark will be thrown across a gap one inch long, between **two sharp points**, not between balls, plates, etc. We always guarantee our coils to give the full spark length. To get best results use storage cells, which are by far the most efficient current. Our Gordon batteries will range next in efficiency. **We do not guarantee** the spark length with **dry cells**, as their current is not steady enough. However, it is understood that they will give excellent results if treated intelligently.

The secondaries of all our coils are wound with enameled wire. Our competitors use bare wire, which leaks and reduces the efficiency of the coil.

> No. CX4360   ¼ in. use 2 type R. E. cells, or 3 dry cells
> No. CIX1087   ½ in. use 2 type R. E. cells, or 4 dry cells
> No. EGK1088   1 in. use 3 type R. E. cells, or 5-6 dry cells
> No. GBK4366   1½ in. use 3-4 type R. E. cells, or 6-7 dry cells
> No. IFK1089   2 in. use 4 type R. E. cells, or 12 dry cells

PRICE LIST

No. CX4360 ¼ inch coil, price.............................. Size 7½x4 5/16x6¼. Shipping weight 4 lbs.	**$ 3.00**
No. CIK1087 ½ inch coil, price.............................. Size 7½x4 5/16x6¼. Shipping weight 5 lbs.	**$ 3.90**
No. EGK1088 1 inch coil, price.............................. Size 7¾x5¾x6¼. Shipping weight 6 lbs.	**$ 5.70**
No. GBK4366 1½ inch coil, price.............................. Size 8½x5¾x6¼. Shipping weight 7 lbs.	**$ 7.20**
No. IFK1089 2 inch coil, price.............................. Size 8½x5¾x6¼. Shipping weight 8 lbs.	**$ 9.60**
No. BCX1090 3 inch coil, price.	**$23.00**
No. CFX1091 4 inch coil, price.	**$36.00**
No. LEX1093 6 inch coil, price..	**$95.00**
No. AAEX1094 8 inch coil, price	**$115.00**

Cabinet style.

Prices of larger coils on application.

Experiments With Spark Coils

Connect two short pieces of wire to the two top binding posts. Make a "spark gap" by leaving a small space between the wire points. If the coil is started a steady stream of sparks will flow between the points. It can be intensified by tightening the thumb screw on the vibrator of the coil. If the "spark gap" is about ¼ inch, a "fire ball" will be observed between the points. If the experiment is continued the positive wire will get white hot and finally fuse at the end. If the wires were copper the fire ball will be green; if of iron, reddish yellow; if of zinc, bluish. To lengthen the spark, attach a metal ball, or metal disc, to the negative pole. The positive pole should have a sharp point. The lengthened spark will not be single; it will tend to branch out.

Another method to greatly lengthen the spark is as follows: Moisten the cover of coil frame between the two wire points with your finger. The spark will at first be thin, but it will enlarge gradually as the

Experiments with Spark Coils
(Continued)

moisture dries. This method lengthens a spark two or three times. A 1-inch spark coil will very often give 2 or 3 inches. The experiment is very interesting. If a thin glass plate is placed in the spark gap, the spark will not be straight, but it will hit around the plate's edge in zigzag form. A very striking experiment is done as follows: Bend two thin iron wires vertically in such a manner that they run parallel. With a little experimenting the right distance to space the wires will be found. The spark will then start at the bottom and run up swiftly in ladder fashion. As soon as it reaches the top it stops, only to recommence at the bottom. It will work automatically for hours, and never fails to attract considerable attention. The sparks also emit a strange noise.

Lengthen the spark gap 4-5 times and strew carbon powder or metal filings between it. The spark will select a route of its own in a peculiar manner. The experiment is greatly beautified in the dark. A small cup of benzine, gunpowder, etc., can be exploded if placed in a spark gap; but of course great precaution is necessary for such experiments. If the flame of a candle is brought near the spark gap, the spark will be drawn into the flame (hot air being a better conductor for the current than cold air). If the candle is blown out and if the wick is touched at once by the spark, it will light up again. If a piece of cardboard is put between the spark gap it will be pierced. The bigger the coil the thicker the cardboard can be. Note the very peculiar hole the spark has made, and compare it with a hole the needle has made. Explanation: The current comes from both sides.

If your friend smokes cigarettes you can play an amusing trick on him. Offer him some of your cigarette paper prepared as follows: Place 10 or 15 leaves on a metal plate to which one wire of the coil leads. Take the other wire (which must be well insulated or you get a shock) and move it all over the surface of the cigarette paper; the more sparks you make in different places the better the trick will turn out. The idea is this: The paper will be pierced with numerous holes,—too fine to be observed,—and when your friend tries to light the cigarette after he carefully rolled it, he will waste a box of matches without being able to get as much as one puff. After trying three or four leaves you can hardly blame him if he commences to say a few things—or he may quit smoking cigarettes altogether. If an old incandescent bulb is connected with one wire, and if the other is grounded, the bulb will emit a greenish light in the dark, as soon as the coil starts working. If one wire has a very fine point and is not too far away from the other wire, a very peculiar and weird discharge will be observed in the dark. If a drop of oil is placed in the spark gap it will be scattered around violently through the spark. If your neighbor's dog has the habit of extracting things from your ash can, lead a well insulated wire to the can, which must stand on a piece of very dry wood. Ground the other wire. When you see the dog standing on his hind legs and leaning against the can, bring your coil in operation. You will never see a more surprised dog in your life, and you can vouch that he will never come near that particular ash can again, even if it should be full of soup bones.

We leave it to the ingenuity of the experimenter to devise new experiments, tricks, etc., and shall be pleased to hear of such, for the benefit of other experimenters.

The most beautiful and startling effects, however, are created by lighting Geissler tubes.* Our smallest coil will light the very biggest tube for hours, and our 1-inch coil will light 8-10 medium tubes simultaneously if connected in series. As all tubes are different in color, the most beautiful effects and designs can be created. In a store window they will stop every passer-by without fail. For parties a weird effect is obtained by suddenly turning out all the lights and operating a single large tube near the ceiling. It will startle the most phlegmatic man, and the ladies will swear they saw a ghost.

IMPORTANT NOTE: If for some reason a coil does not give the right spark length, bend the vibrator spring a little back (towards the thumb screw). The stiffer a vibrator works the better the spark will be.

Geissler tubes, being imported from Germany cannot be obtained at present.

How to Photograph Electrical Discharges

By H. GERNSBACK,

President Electro Importing Company,

Editor "The Electrical Experimenter."

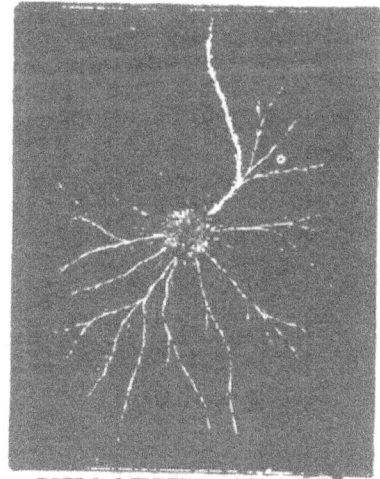

NEGATIVE SPARK

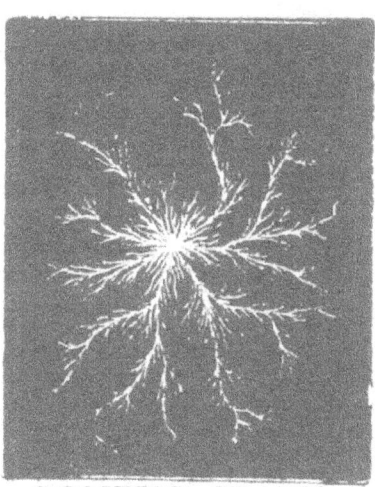

POSITIVE SPARK

The following most extraordinary experiments, which, of course, can be varied hundredfold, can be performed with any of our coils, the ¼ inch one included. The ordinary photographic plate is used for all the experiments, its size depending on the objects.

It is self-evident that such experiments must be conducted in a dark room or in a room lighted only with a ruby (red) photographic lamp. Any white light will spoil the plates instantly. After exposing, the plates must of course be developed, which you can easily do yourself, or else re-wrap it in its black paper and have a photographer develop and print the negative.

Take a small bottle with wide mouth and fill it half full with very dry and pure starch powder, sulphur flour, or with talcum powder. Over the mouth place a thin piece of gauze to act as a fine sieve. Tie the gauze around the neck of the bottle with a string. Take the photographic plate and place it (with the coated side upwards) on a metal plate, or a piece of stiff sheet iron, tin, etc. Connect the metal plate with one of the secondary posts of your coil.

Sieve a thin film of the above-named powder over the photographic plate. Now place a very fine metal point (big needle or pin) in the middle of the prepared plate. Connect the point with the other post of the coil, and make one spark. If the duration of the discharge is longer than one second, the results will not be as nice. The plate, after the powder has been wiped off completely with a soft cloth, is ready for developing.

Different results will be obtained if the polarity of the metal point is changed.

Fig. No. 1 (made with a ½ inch coil) had the point connected with the negative pole. The point of Fig. No. 2 (same coil) was positive. Note the white centre ball of Fig. No. 1.

The most beautiful symmetrical and other designs can be made as follows: Cut a pattern (such as a star, your initials, etc.), in a piece of cardboard and place this on the photographic plate. Now sieve the powder over the pattern, and when this is removed the design alone will show on the plate. Place the metal point in the centre of the design and make a spark as explained above. Of course, no two photographs will ever be alike, and the greatest surprises are experienced by the creation of new designs, branchings, etc. The result of experiments in this field are most interesting and grateful.

Penn Yan, N. Y.

Dear Sirs:—

Received your information that I asked of you and thank you a thousand times for the same. I would not take $10.00 FOR THE COIL I RECEIVED OF YOU AT $4.75. IT WILL JUMP AN INCH ON FOUR COLUMBIA DRY CELLS.

Yours respectfully, CHAS. CARRY.

90 MILES WITH AN E. I. Co. COIL

WORLD'S RECORD BROKEN
SPECIAL AEROGRAM SENT AS TEST OF NEW AERIAL

On the 17th of April, 1910 at 5:15 P. M., Ray Newby operating for the school of Wireless, broke all previous records for the most efficient transmission, sending to Mare Island a distance of nearly **70 MILES** from San Jose, with an expenditure of not more than **15 WATTS OF ENERGY.** Further tests were carried on the following day the Farallon Island registering interference showing that the earth's potential had been disturbed at this distance ABOUT 90 MILES.

The current was supplied by a small storage battery actuating a **1 INCH E. I. CO. COIL.** The spark gap used was a small E. I. Co. Zinc gap set at about 3 millimeters. The (apparent) wave length was about 500 meters, the frequency 600,000 cycles per second. Following is a copy of the message sent at that time:

1 FN RA Ck Dh Fn

REAR ADMIRAL H. OSTERHAUS,

Mare Island, Calif.

Congratulations from the Soldier boy whose gun your honored Father shot in rifle pits at Spanish Forts.

W. G. HAWLEY, Postmaster.

3/26/10

SCHOOL OF WIRELESS

Garden City Bank Building
San Jose, Calif., June 23, 1910.

The Electro Importing Co.,
233 Fulton St., New York City.

Gentlemen:——

We have transmitted messages from our station to the Government Stations at Mare Island and the Farallon Island, A DISTANCE OF 90 MILES, and also to the U. W. T. Co. Station in the Crocker Tract, San Francisco, USING ONE OF YOUR ONE INCH (1") coils and a small portable storage battery. We have made some of our best of the above tests at midday USING THE SAME ONE INCH COIL. We have given wireless phone concerts to amateur wireless men throughout the Santa Clara Valley, USING THE SAME ONE INCH COIL and an Ericsson Dust Transmitter.

(Signed) CHAS. D. HERROLD.

STATE OF CALIFORNIA,
ss. :
COUNTY OF SANTA CLARA.

I, Chas. D. Herrold, being duly sworn deposes and says that the above matter to which my signature is attached is true to my best knowledge and belief.

CHAS. D. HERROLD.

Subscribed and sworn to before me this 24th day of June, 1910.

WESLEY PIAPA,

Notary Public in and for County of Santa Clara, State of California.

San Jose, Calif., June 23, 1910.

TO WHOM IT MAY CONCERN:

Mr. Raymond Newby and I USING AN E. I. CO.'S ONE INCH COIL, and E. I. Co.'s zinc gap set at about 1/16 inch, and the antenna of the School of Wireless in the Garden City Bank Bldg. called up operator RH of Mare Island Station getting an immediate response. He gave us time from the standard clock and told us that we came in strong.

I have also heard Operator Newby talk to PH, the big U. W. T. Co.'s Station in the Crocker Tract, San Francisco, and also heard the Farallon Island Station tell us "Keep out." In every case we used the **SAME ONE INCH COIL.** I have also, using the same set, talked with PH myself two different times.

(Signed) THAD STEVENS.

STATE OF CALIFORNIA,
ss. :
SANTA CLARA CO.

I, THAD STEVENS, being duly sworn, deposes and says that the above facts as therein set forth are true to my best knowledge and belief.

THAD STEVENS.

Subscribed and sworn to before me this 25th day of June, 1910.

WESLEY PIAPA,

Notary Public in and for County of Santa Clara, State of California.

San Jose, Calif., June 23, 1910.

TO WHOM IT MAY CONCERN:

I transmitted the message from Major Hawley to Admiral Osterhaus using a one inch E. I. Co.'s (1") COIL actuated by a small portable storage battery. The energy used was LESS THAN 15 WATTS, the distance covered being that between San Jose, Calif., and Mare Island. I used the system of the School of Wireless, designed and built by Chas. D. Herrold, the Electrical Engineer for the Company. I also talked with Operator Ludwig of the Farallon Islands and PH United Wireless Station in San Francisco using the same hook-up and using the same one inch coil. Mr. Thad Stevens was present at several of the tests and himself talked with the last named station.

(Signed) RAY NEWBY.

STATE OF CALIFORNIA,
ss. :
COUNTY OF SANTA CLARA.

I, Ray Newby, being duly sworn deposes and says that the above is true to my best knowledge and belief.

(Signed) RAY NEWBY.

Subscribed and sworn to before me this 24th day of June, 1910.

WESLEY PIAPA,

Notary Public in and for County of Santa Clara, State of California.

CHEMISTRY DEPARTMENT

We have had lately so many demands for real practical chemical outfits and appliances that we have decided to put in a line of this useful apparata which we present herewith to our friends. It is understood, of course, that pursuant to our policy to give our customers only the best in the market, our chemical department is up to this standard and besides the prices are so reasonable that everybody can afford to buy one of these complete sets.

Our Chemical Set is endorsed by Chemical Experts, Government Officials, and Professors of Chemistry in leading schools and colleges.

There is hardly a large manufacturing firm in the country that does not employ a chemist, and they are looking around trying to find more chemists, but the supply is **far short** of the demand. Very high fees are paid to good chemists, and here is your opportunity waiting for you.

Of course, our Chemical Set cannot contain all the different apparata necessary for a more advanced study of Chemistry and we have therefore listed in the following pages a quantity of Laboratory Glassware, Instruments, Reagents and Chemicals to supplement our Laboratory Outfit.

THE E. I. CO. CHEMICAL LABORATORY

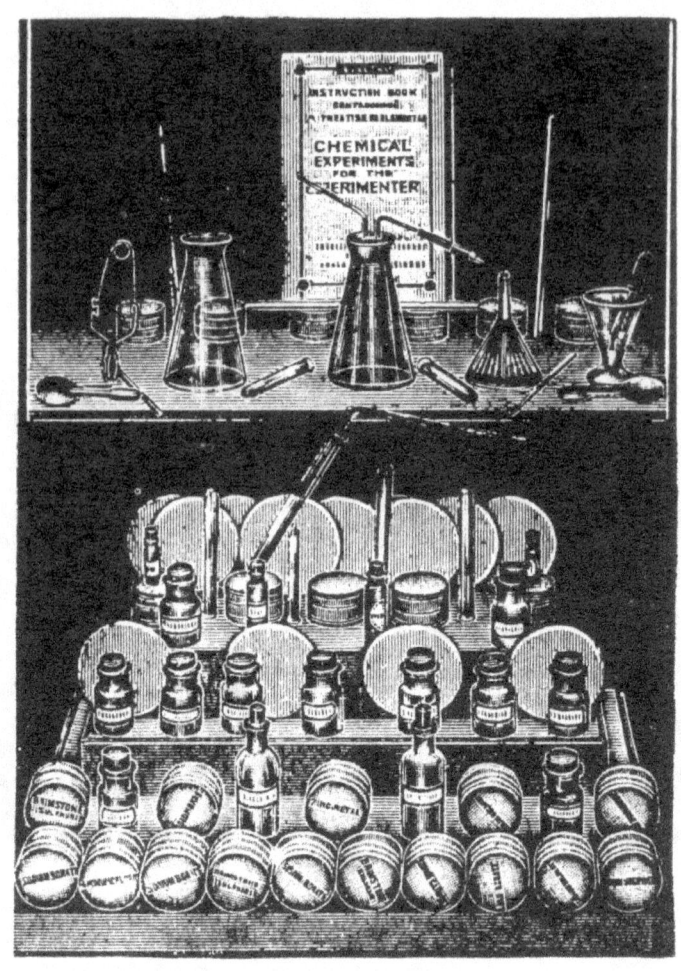

No. 4889   Chemical Laboratory

44 CHEMICALS
18 APPARATA
INSTRUCTION BOOK WITH 100 EXPERIMENTS

PRICE
$5.00

Shipping Weight 10 lbs.
CAN BE SHIPPED BY EXPRESS ONLY

We present herewith to our friends our new **E. I. Co. Chemical Laboratory** which contains **real** chemicals and apparata to perform **real** chemical experiments. This outfit is not a toy, put up merely to amuse, but a practical laboratory set, with all the chemicals, apparata and reagents necessary to perform real work and to teach the beginner all the secrets of inorganic chemistry. With this outfit we give free a book containing a **Treatise in Elementary Chemistry**, useful data and **recipes, and 100 instructive and amusing experiments**.

Chemical Laboratory
(Continued)

The chemicals furnished are all technically pure and put up in appropriate wooden boxes and glass bottles, and there is a sufficient quantity to make dozens of experiments with each. The apparata are of standard laboratory size and quality.

Although all chemicals have nearly doubled in price, we have decided not to raise the price for this outfit for the present.

Read the list of chemicals and apparata and look at the actual photograph of the outfit herewith.

Description of the Outfit

It contains the following 44 chemicals:

Ammonium Chloride	Zinc, Metallic	Chloride of Zinc
Alum	Sodium Bicarbonate	Copper Sulphate
Antimony	Sodium Sulphate	Glycerol
Boracic Acid	Sodium Chloride	Iron Chloride
Charcoal	Calcium Sulphate	Calcium Oxide
Sodium Nitrate	Barium Chloride	Stannous Chloride
Sodium Carbonate	Lead Acetate	Nickel Chloride
Sodium Borate	Ferrous Sulphate	Hydrochloric Acid
Sodium Sulphite	Nickel Sulphate	Sulphuric Acid
Manganese Dioxide	Sodium Phosphate	Iodine
Oxalic Acid	Zinc Carbonate	Mercury, Metallic
Brimstone	Ammonium Sulphate	Tin, Metallic
Iron Oxide	Ammonium Carbonate	Litmus Paper
Sulphate of Zinc	Ammonium Aqua	Ferrous Sulphide
Magnesia Carbonate	Calcium Chloride	

The following apparata are furnished:

One Standard Washbottle
One Conical Glass Measure
One Erlenmeyer Flask
One Glass Funnel
One Delivery Tube
Six Assorted Test-Tubes
One Test-Tube Holder
Ten Sheets of Filter Paper

One Glass Dropper
One Spoon Measure
One Spirit Lamp
Glass Tubing
One book containing Treatise on Elementary Chemistry and 100 Chemical Experiments to be performed with this outfit.

No. 4889 E. I. Co. Chemical Laboratory, as described.......... **$5.00**
Shipping weight 10 lbs.

Laboratory Glassware and Apparatus.

The articles described below are of standard shape and size. The glassware is in the highest degree resistant to the action of chemical reagents and to temperature changes, and thoroughly tested before leaving our stock rooms.

Owing to the breakable nature of this material we recommend it sent by express or freight.

No. DK2092

Glass Jars and Bottles.

No. DE2090 Round Specimen Jar. Made of clearest flint glass. Size 4¾ x 9½ in. Shipping weight 4 lbs. **$0.45**

No. DK2091 Round Specimen Jar. Same as above. Size 4½x6½ in. Shipping weight 3 lbs. **$0.40**

No. DK2092 Square Glass Tank. Size 3¾x4x5. Shipping weight 3 lbs. **$0.40**

No. AH2093 Large Glass Bottle with Cork Stopper. Size 3¾ in. high, diameter of bottom 1⅞ in., diameter of cork 1¼ in. Shipping weight 1 lb. **$0.18**

No. AE2094 Small Glass Bottle with Cork Stopper. Size 2 in. high, diameter of bottom 1¼ in., diameter of cork 1⅛ in. Shipping weight 4 oz. **$0.15**

No. CK2108

No. CK2108 Chemical Flask. Flat Bottom Lip Finish. Containing 4 oz. (120 c.c.).. **$0.30**

No. CE2109 Same, containing ½ pint (250 c.c.). **$0.35**

Shipping weight 2 lbs.

No. CK2110 Chemical Flask. Round bottom, lip finish, containing 4 oz. (120 c.c.)... **$0.30**

No. CE2111 Same, containing ½ pint (250 c.c.) **$0.35**

Shipping weight 2 lbs.

No. CK2110

No. CK2112

No. CK2112   Erlenmeyer Flask.   4 oz. (120 c.c.).   **$0.30**

No. CE2113   Same, ½ pint (250 c.c.)............   **$0.35**

Shipping weight 2 lbs.

Glass Funnel. A cheaper grade of funnel made of pressed glass, useful where the more expensive one, shown elsewhere, is not needed. Dimensions: Length 4 inches, diameter 2½ inches.

No. 2182 Funnel, as described...............................   **$0.15**

Shipping weight 1 lb.

No. 2183 Conical Glass Graduate. Made of pressed glass. Graduations not etched as in the more expensive ones, but fairly accurate. Measure indication: 2 ounces, divided in 8th, 4th and ½ oz. Dimensions: Height 3½ in., diameter 2 in.

No. 2183 Graduate, as described **$0.20**

Shipping weight 1 lb.

Desiccating Jar

No. AX2127   With ground lid, 3⅞ inches inside diameter **$1.00**

Shipping weight 5 lbs.

Watch Glasses

No. BK2128   Syracuse pattern.   Outside diameter 65 mm.; inside diameter 50mm.; depth 10 mm. Bevel ground to form surface for writing.   Each **$0.20**

Shipping weight 1 lb.

No. AX2127

No. E2129

Watch Glasses

No. E2129   Standard pattern, 2 in. diameter   **$0.06**

No. AK2130   Standard pattern, 4 in. diameter **$0.10**

No. AF2131   Standard pattern, 6 in. diameter **$0.16**

Shipping weight 1 lb. each.

No. AX2133

Wash Bottle

No. AX2133 Complete with glass tubes and rubber stopper; flexible exit tube. 1 pint capacity **$1.00**

Shipping weight 3 lbs.

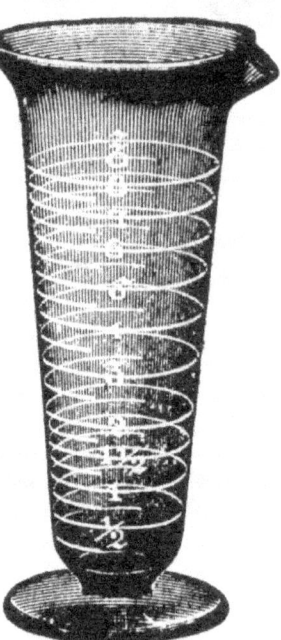

Graduate

No. AX2134 Cone shape. On foot with lip. Hand engraved, clear-cut graduations. Duplex scale, fluid measure on one side and metric measure on the other. To deliver 2 ounces (60 c.c.) **$1.00**

Shipping weight 1 lb.

No. ABK2135 GRADUATE. Same. To deliver 4 ounces (120 c.c.)............ **$1.20**

Shipping weight 2 lbs.

No. AFE2136 GRADUATE. Same. To deliver 8 ounces (250 c.c.)............ **$1.65**

Shipping weight 3 lbs.

No. AX2134

No. ADK2137

Graduate

No. ADK2137 Cylindrical shape, otherwise the same as above.

To deliver 50 c.c. graduated in 1 c.c. each...... **$1.40**

Shipping weight 1 lb.

Evaporating Dishes

No. EK2140 Best porcelain, 3½ inches diameter. Contents 2 oz. Each **$0.50**

No. FK2141 Same. 4¼ inches diameter, contents, 4 oz. Each... **$0.60**

No. GE2142 Same. 5¼ inches diameter, contents, 7 oz. Each.. **$0.75**

No. AX2143 Same. 6½ inches diameter, contents, 1 pint. Each.. **$1.00**

Shipping weight 2 lbs. each.

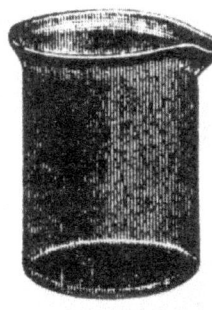

No. IE2115

Beaker Glasses

No. IE2115 Wide with Pour Out, Griffin's form. **Nested** in assortment of 5 Beakers of 1, 2, 3, 4, and 6 oz. contents. Price per set **$0.95**

Shipping weight 3 lbs.

Test Tubes

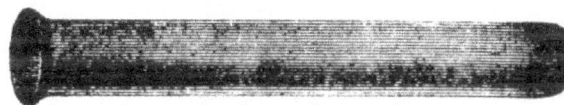

No. CC2116

No. CC2116 Non-corrosive under ordinary conditions and resistant to sudden changes in temperature. Nested in assortment of 3, 4, 5, 6, 7 and 8 inches long. Price per set.......... **$0.33**

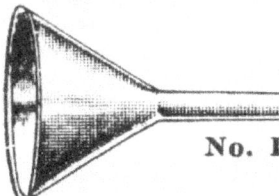

No. BK2117

Funnels

Long stem. Body made uniformly at an angle of 60°. Stem straight to facilitate rapid filtration.

No. BK2117 Diameter 1 inch (2.5 c.m.)..................... **$0.20**

No. CF2118 Diameter 2 inch (5 c.m.)..................... **$0.36**

No. EE2119 Diameter 4 inch (10 c.m.)..................... **$0.55**

Shipping weight 2 lbs. each.

  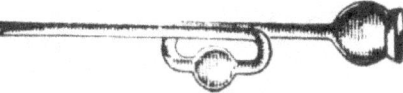

No. DK2120   No. FE2121   No. GE2122

Thistle-Tubes and Safety Funnels

Best Nonsol Glass

No. DK2120 Straight stem **$0.40**

No. FE2121 One loop **$0.65**

No GE2122 One loop and bulb **$0.75**

Shipping weight, each, 2 lbs.

No. FK2144

Glass Mortar and Pestle

No. FK2144 Contents 4 oz. **$0.60**

Shipping weight 1 lb.

No. GE2145 Same, contents 8 oz........ **$0.75**

Shipping weight 2 lbs.

Tube Connections—Glass

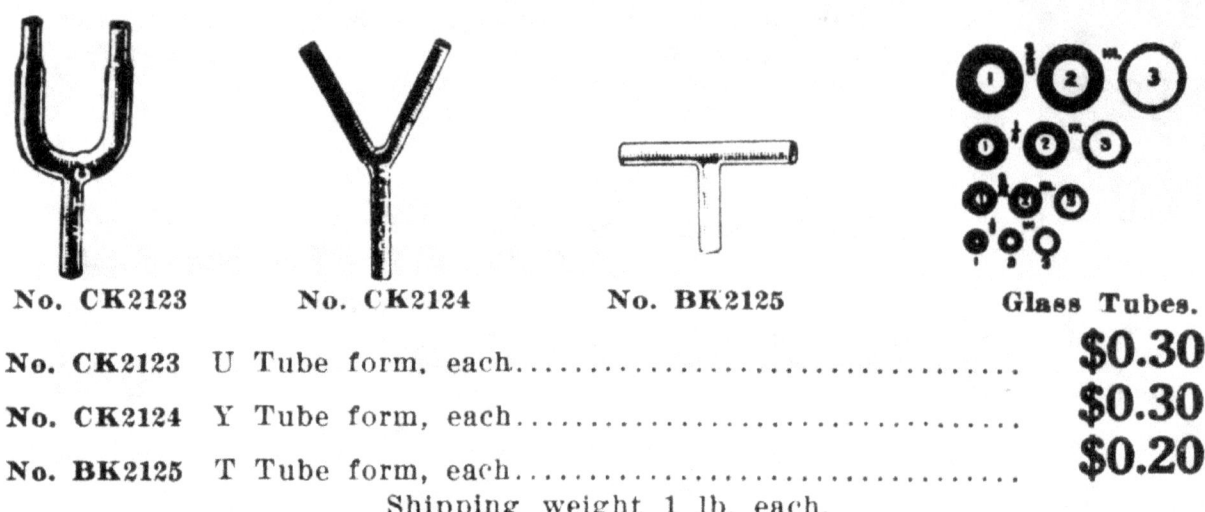

No. CK2123   No. CK2124   No. BK2125   Glass Tubes.

No. CK2123 U Tube form, each..................................		**$0.30**
No. CK2124 Y Tube form, each..................................		**$0.30**
No. BK2125 T Tube form, each..................................		**$0.20**

Shipping weight 1 lb. each.

GLASS TUBING. This tubing is of the best quality of glass, and will stand heating and bending into shapes without breaking. We cannot sell less than one foot nor longer lengths than two feet.

Catalogue No.	Outside Dia.	Approximate Inside Dia.	Price per Foot
No. I6290	$\frac{1}{8}$ in.	$\frac{1}{16}$ in.	$0.09
No. AB6291	$\frac{3}{16}$ in.	$\frac{1}{16}$ in.	0.12
No. AE6292	$\frac{1}{4}$ in.	$\frac{1}{8}$ in.	0.15
No. CK6293	$\frac{3}{8}$ in.	$\frac{3}{16}$ in.	0.30
No. DE6294	$\frac{1}{2}$ in.	$\frac{5}{16}$ in.	0.45
No. FE6295	$\frac{3}{4}$ in.	$\frac{7}{16}$ in.	0.65

Shipping weight 2 to 3 lbs. a foot, owing to the breakable nature of this material.

GLASS RODS. Our glass rods come only in full foot and 2 feet lengths. Under no circumstances can we sell shorter lengths. Our glass rods are especially recommended for electrical work, for high tension Tesla experiments, static machines, etc. Warranted not to contain lead or conducting salts.

Catalogue No.	Diameter	Price per Foot
No. AK2105	$\frac{1}{8}$ in.	$0.10
No. AE2106	$\frac{1}{4}$ in.	0.15
No. BG2107	$\frac{1}{2}$ in.	0.27

Shipping weight 2 to 3 lbs. a foot, owing to the breakable nature of this material.

SOFT RUBBER TUBING. A good grade of Para Rubber suitable for all experiments. It will stand hot or cold water and many chemicals without drying and breaking as experienced in many cheap grades. White Rubber. heavy walls.

Catalogue No.	Inside Diameter	Price per Foot
No. AE2102	$\frac{1}{8}$ in.	$0.15
No. BE2103	$\frac{3}{16}$ in.	0.25
No. CE2104	$\frac{1}{4}$ in.	0.35

Shipping weight per foot 1 lb.

Reagent Bottles

These bottles are made of glass, free from lead, zinc or other metallic flux. Flat hood, glass stoppers, smooth bottoms, narrow mouth.

No. BDE2148 Contents 1 oz., height 3⅜ in., per doz. **$2.45**

No. CFK2149 Contents ¼ pint, height 4⅞ in., per doz. **$3.60**

No. EX2150 Contents 1 pint, height 7⅛ in., per doz. **$5.00**

No. BDE2148

No. BEK2151 Same. Wide mouth. Contents 1 oz., height 3⅜ in., per doz................. **$2.50**

No. CGE2152 Same. Wide mouth. Contents ¼ pint, height 4⅝ in., per doz................. **$3.70**

No. EAK2153 Same. Wide mouth. Contents 1 pint, height 6¾ in., per doz............... **$5.10**

No. BEK2151

Crucibles

No. GE2154 Conical form. Sand, 6 in nest, height of largest 5½ inches, width on top 4½ inches. Per nest of 6............... **$0.75**

Shipping weight 3 lbs.

No. GE2154

Filter Paper

Best quality. Unsurpassed in strength, uniformity of texture and clear rapid filtering. Cut round, white grade.

No. DE2155 Diameter 6 inches, round, per pack of hundred..... **$0.40**

No. EE2156 Diameter 8 inches, round, per pack of hundred..... **$0.55**

No. GE2157 Diameter 10 inches, round, per pack of hundred...... **$0.75**

Shipping weight 1 lb. per pack.

Engraved Stem Chemical Thermometer

No. AIK2181 Fahrenheit and centigrade scale. Length 14 inches, ¼ in. diameter. Scale range 30° to 400° F. Subdivision 2° F. and 1° C. Each, in turned wood box........................... **$1.90**

Shipping weight 2 lbs.

No. BEK2158

Water Bath

No. BEK2158 Polished copper, tin-lined, with concentric copper rings, cover and steam escape. Diameter 5 inches, 4 rings... **$2.50**

Shipping weight 3 lbs.

Sand Bath

No. DK2159 Deep form, best iron. Diameter 6 inches **$0.40**

Shipping weight 2 lbs.

No. DK2159

No. AEK2160

Tripod

No. AEK2160 Galvanized iron, 7⅜ inches high. Ring 4⅝ inches diameter. With adjustable lamp bracket.......... **$1.50**

Shipping weight 3 lbs.

Wire Gauze

$0.25

No. BE2161 Iron, for use on tripods. 6 inches square........... Shipping weight 1 lb.

Blowpipe

No. CK2162 Black's model. Conical shape, of Japanned tin, with detach able brass tip. **$0.30** Shipping weight 2 lbs.

No. CK2162

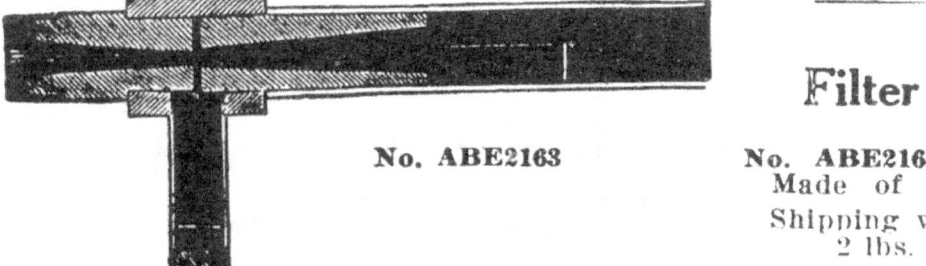

No. ABE2163

Filter Pump

No. ABE2163 (Aspirator). Made of brass. **$1.25** Shipping weight 2 lbs.

Spirit Lamp and Bunsen Burner, see page 159.

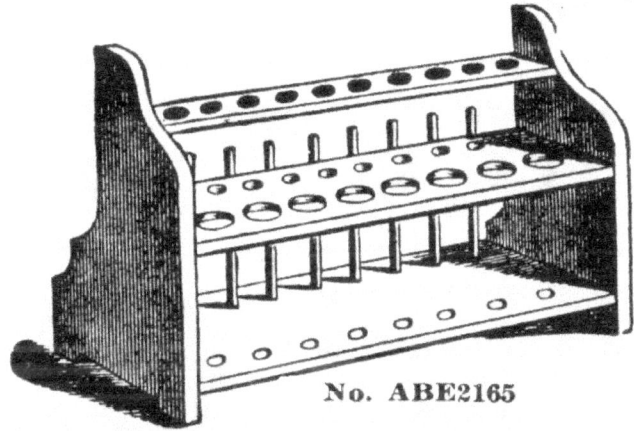

No. ABE2165

Test Tube Rack

No. ABE2165 Wood; for 18 Tubes, with 8 pins for draining. Each **$1.25**
Shipping weight 5 lbs.

Test Tube Holder

No. CK2166 Wood; with wire spring. Each **$0.30**

No. CK2166

TEST TUBE HOLDER

No. CK2167 Wire, spring brass, nickel plated....... **$0.30**
Shipping weight 1 lb., each kind

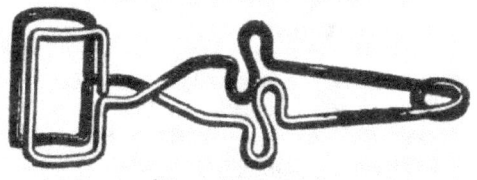

No. CK2167

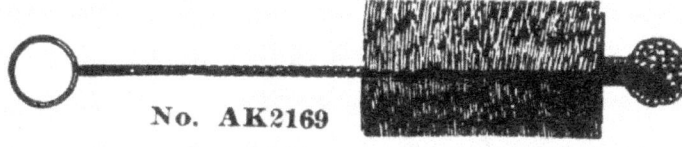

No. AK2169

Test Tube Brush

No. AK2169 On tinned wire. Shipping weight 4 oz. **$0.10**

Burrette Clamp

No. GE2170 Iron, with check nut to adjust position. Each **$0.75**
Shipping weight 1 lb.

No. GE2170

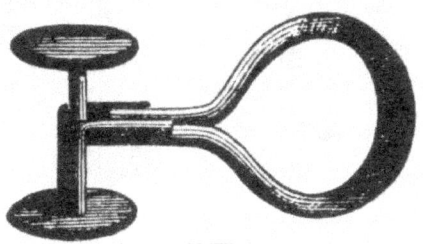

No. BK2171

Spring Pinch-Cock

No. BK2171 Mohr's pattern, brass nickel plated. Each............. **$0.20**
Shipping weight 4 oz.

Clay Triangles

No. AE2168 2 sizes: 2 and 3 inches. State which size in ordering. Each....... **$0.15**
Shipping weight, each, 4 oz.

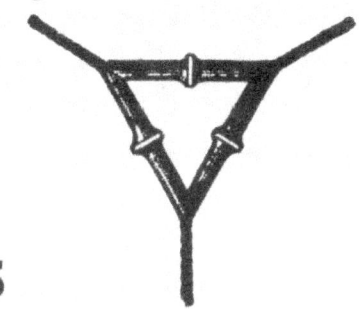

No. AE2168

Filter Stand

No. AGE2164 Iron with 3 rings.................................... **$1.75**

Shipping weight 4 lbs.

Reagents and Chemicals.

We do not charge packing on dry chemicals. Bottles, to ship liquids, are included in price. Bottles to contain acid are provided with glass stoppers.

Acid Acetic U.S.P.	$HC_2H_3O_2$	Liquid	lb.	$0.50
Acid Hydrochloric (Muriatic).......	HCl	Liquid	lb.	0.50
Acid Nitric ch. pure.................	HNO_3	Liquid	lb.	...
Acid Oxalic. tech.............	$H_2C_2O_4$	Solid	½ lb.	0.80
Acid Sulphuric ch. pure...........	H_2SO_4	Liquid	lb.	0.45
Alcohol denatured		Liquid	lb.	0.55
Alcohol, Wood, refined 95% (Methyl)	CH_3OH	Liquid	lb.	0.50
Ammonia, concentr. (Hydroxide)....	NH_4OH	Liquid	lb.	0.50
Ammonium Carbonate, tech.	$(NH_4)_2CO_3$	Solid	lb.	0.50
Ammonium Chloride (Sal) tech.....	NH_4Cl	Solid	lb.	0.60
Barium Chloride, tech.	$BaCl_2$	Solid	lb.	0.45
Calcium Chloride	$CaCl_2$	Solid	lb.	0.25
Calcium Sulphate	$CaSO_4$	Solid	lb.	0.20
Chloroform U.S.P.	$CHCl_3$	Liquid	½ lb.	0.80
Cupric Sulphate, tech.............	$CuSO_4$	Solid	lb.	0.30
Ether	$(C_2H_5)_2O$	Liquid	lb.	0.65
Formaldehyde, tech. (Formalin)....	CH_2O_2	Liquid	lb.	0.60
Glycerine ch. pure	$C_3H_5(OH)_3$	Liquid	½ lb.	0.75
Hydrogen Peroxide, tech.	H_2O_2	Liquid	lb.	0.45
Iodine, Resublimed	I.	Solid	oz.	0.65
Iron Chloride (Ferric)	$FeCl_3$	Solid	lb.	0.45
Iron Sulphate, tech.	$FeSO_4$	Solid	lb.	0.18
Lead Acetate, tech.	$Pb(C_2H_3O_2)_2$	Solid	lb.	0.45
Lead Nitrate, tech.	$Pb(NO_3)_2$	Solid	lb.	...
Litmus Paper, blue.................			book @	0.10
Litmus Paper, red.................			book @	0.10
Manganese Dioxide, tech.	MnO_2	Solid	lb.	0.45
Mercuric Chloride	$HgCl_2$	Solid	½ lb.	...
Mercury Metallic, tech.	Hg.	(Quicksilver)	½ lb.	0.95
Nickel Chloride	$NiCl_2$	Solid	½ lb.	0.60
Potassium Bromide U.S.P...........	K Br	Solid	½ lb.	1.50
Potassium Carbonate, tech.	K_2CO_3	Solid	½ lb.	...
Potassium Chlorate, tech.	$KClO_3$	Solid	½ lb.	...
Potassium Cyanide	KCN	Solid	½ lb.	...
Potassium Hydroxide (Caustic)	KHO	Solid	½ lb.	...
Potassium Nitrate	KNO_3	Solid	½ lb.	...
Potassium Permanganate	$KMnO_4$	Solid	oz.	...
Silver Nitrate cryst.	$AgNO_3$	Solid	oz.	...
Sodium Hydroxide (caustic)	NaOH	Solid	lb.	1.00
Sulphur (Brimstone)	S	Solid	lb.	0.25
Stannous Chloride (Tin)..........	$SnCl_2$	Solid	½ lb.	0.75
Zinc (Mossy) Metal	ZN	Solid	½ lb.	0.25

All acids have to be shipped by express.

NOTICE: Above Chemicals are put up in Standard packages and bottles and cannot be sold in smaller quantities. Special price for bigger quantities upon request. The chemicals marked (...) cannot be sold during the war.

Omnigraphs
(AUTOMATIC TELEGRAPH MACHINES)

This wonderful instrument has been produced to fill a gap that has existed for years. It is the only apparatus that will automatically teach you telegraphy without a teacher. The Omnigraph teaches you telegraphy, as well as **Wireless, Continental and Morse** code at your own home at a ridiculously low cost. Our instrument will positively teach you **better than any teacher could, and in less time.** It actually takes the place of an expert and **will send you messages at any speed you desire.** You can send a single letter continuously or a short message and gradually make the message more difficult. First you learn all the letters, then you read a short message, then you can reverse the dial and the Omnigraph will send part words and part letters. We furnish a large assortment of dials which will be found listed below.

The Omnigraph works perfectly on any line or with any instrument.

WIRELESS—If you have no one to teach you Wireless, the Omnigraph will do it with astonishing rapidity. By connecting our RADIOTONE or our No. EK965 or our No. DK950 buzzers with the Omnigraph, you will get a close imitation of a wireless message, and there is positively nothing like it made. **You can also connect your wireless 'phones across the electromagnets of the buzzer, putting a small condenser between 'phones and magnets** and you will hear then a message that **positively cannot be distinguished from a real wireless message.**

All our Omnigraphs operate sounders, relays, bells, our RADIOTONE, buzzers, wireless sending apparatus, etc., etc. Our records being made of metal are everlasting and cannot wear out.

Omnigraph No. AFX2777

No. AFX2777

This is our cheapest instrument and it is one of the most ingenious telegraphic machines ever invented. A masterpiece in all respects.

Sends absolutely perfect at any speed from 10 to 100 words per minute. You can change the message in the fraction of a second even while the machine is running. You can send the same message continuously or a 5th part repeatedly.

This Omnigraph is provided with five movable message changers. Each dial is divided into five equal parts making the dials so far as changing the message is concerned **equal to 25 dials.** Starting the record when the dials are placed on the spindle from "A" to "E" having the dial marked "A" on top, it sends a comprehensive message of 50 words and before repeating the first dial it is equal to 100 words.

To change the message you move No. 1 message changer (the lever under the letters) in, so that it does not engage with the star wheel and move No. 2 out, this will transpose ten words. By making this change with the five different message changers it will transpose 50 words. Now by putting 1 and 2 so that they engage with the star wheel, this will make 20 changes, and by making this change five times using different message changers each time, you make 100 changes. This same kind of manipulation can be done with three message changers, or four and with five, which will make the same proportion of changes in the message. Now remove the dials, shuffle them up, replace them on the spindle and you will begin all over again with new messages so far as practice is concerned. Just apply the rule of permutation to this Omnigraph and you will agree with us that you can send **thousands of new messages.**

If your memory is good we guarantee that you will master telegraphy in one month, practicing each day. One set of 5 dials furnished. Size 11x6x5 inches.

No. AFX2777 **Omnigraph as described.** Shipping weight 7 lbs.... **$16.00**

DIAL LIST FOR OMNIGRAPH NO. AFX2777

No. 7200 to 7214, Fifteen Dials (3 sets of five Dials) Commercial Messages.
No. 7215 to 7229, Fifteen Dials (3 sets of five Dials) Commercial Messages.
No. 7230 to 7244, Fifteen Dials (3 sets of five Dials Railroad Messages.
No. 7245 to 7259, Fifteen Dials (3 sets of five Dials) Continental Code Messages.
No. 7260 to 7274, Fifteen Dials (3 sets of five Dials) Continental Code Messages.

Fifteen Dials form one comprehensive message.

Each set of five Dials, $1.00. Shipping weight 1 lb.

Omnigraph No. BCX6777

This is the largest and newest style made and a great number of Telegraph schools throughout the country have purchased this style and use it to send to their students in preference to hand sending. **The United States Government examines all applicants for wireless licenses, as to their ability to receive messages, by means of this machine.**

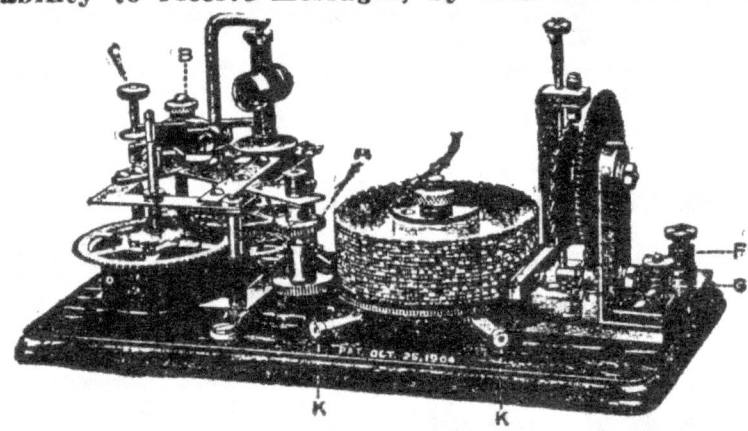

No. BCX6777

The new Omnigraph No. BCX6777 has 75 dials, as each of the 15 dials is divided into 5 equal parts. The message is cut in the aluminum, and the space between words is exactly at the end of each section. **There are 5 movable message changers which allow you not only to change from one dial to another, but to change from one word to another, automatically.** Each dial can be placed in 5 different positions on the spindle, the changes being made more quickly than you can write them.

You can have the message easy or difficult. You can send what is on each dial continuously, and both see and hear what you send. **You can devote all your time to receiving, as the Omnigraph is always ready, and you waste no time in changing messages or in getting the machine ready to send.**

The new No. BCX6777 is finely finished, nickel plated, and will last a lifetime. The Omnigraph is run by spring motor. The spring motor will run it at the rate of 20 words per minute, ¾ of an hour with one winding. It takes only ½ minute to rewind.

Start the motor and the instrument will send you perfectly, a message of 150 words, including all punctuation marks, numerals and characters, at any speed you desire. Receive this message until you can anticipate what it is sending. To change the message, move No. 1 message changer in and move No. 2 out. After receiving this message with the 30 changes, move No. 2 in and 3 out, then 3 in and 4 out, 4 in and 5 out. Then 1 and 2 out, 1 in and 2 and 3 out, then 3 and 4 out, then 4 and 5 out. Then 1, 2 and 3 out. Then 2, 3 and 4 out, then 3, 4 and 5 out, then 4, 5 and 1 out. Then 1, 2, 3, 4 and 5 out. In fact, you can make innumerable combinations with these message changers. With these changes you will find that you will require at least 10 days' practice before you will be able to anticipate at all. When you find you do anticipate just remove the dials from the spindle, shuffle them up and replace them haphazard and begin all over again. After you have used the instrument for months, you will find that it is just as impossible to anticipate as it was the first time you received from this new machine.

Just take pencil and paper and figure out the number of different arrangements you can make with these 15 dials, and you will readily agree with us that the OMNIGRAPH will send thousands of new messages.

If you should undertake to record on paper tape all the new messages this OMNIGRAPH will send, you would need a strip long enough to reach around the globe.

No. BCX6777 Omnigraph, as described.......................... **$23.00**

Shipping weight 10 lbs.

List of Dials available in both Continental (Wireless) Code and the American Morse Code.

CONTINENTAL (Wireless) CODE

Set 1. Each letter of the alphabet, repeated three times, and the numerals repeated three times, with exaggerated spacing between each character. Also all conventional signs and punctuation marks, each character repeated twice, with exaggerated spacing. Break, circumflexion, accented e, attention, end of message, dotted a, bar indicating fraction, wait, understand, period, comma, semicolon, question, exclamation, hyphen, apostrophe, parenthesis, quotation, colon, transmission finished, error, distress call.

Set 2. Short words with letter spacing equal to 10 dots and word spacing equal to 18 dots. Each word repeated three times. The dials contain words as follows: can, eat, fat, him, ice, jet, sit, rain, pack, arms, etc., etc.

Set 3. Similar to Set No. 2, but with an entirely different set of short words.

Set 4. Four-letter code. Letter spacing equal to 5 dots and word spacing irregular. The 15 dials comprising this set contain 75 four-letter code characters.

Set 7. This set contains a straight comprehensive message of about 150 words, with numerals and punctuation. Spacing regular.

Set 9. Same as Set 7, but with an entirely different message.

AMERICAN MORSE CODE DIALS

Set 5. This set of 15 dials contains a straight comprehensive message of about 150 words, including numerals and punctuation. Commercial message with regular spacing.

Set 6. Railroad dials. Message of 150 words, including numerals and punctuation, adapted for practice in Railroad Telegraphy. Regular spacing.

Set 8. A straight commercial message, including numerals and punctuation, with regular spacing, of approximately 150 words in the American Morse Code.

Please order by number.

Set of 15 Dials in either the Continental or Morse Code for use with the Omnigraph, described above $3.00

The "Electro" Practice Set

This standard practice set will enable anybody to learn the sending of messages of Wireless or Morse Code in a very short time. It is a medium priced apparatus, built on the same principles as the most expensive sets.

It consists of a mahogany finished base of

BEK1140

6¾x4¼ inches, upon which are mounted a telegraph key of standard pattern, a buzzer and three binding posts. The key and buzzer have silver contacts, and as the key has the "click" like a sounder, the set can be used for learning to send messages in the American Morse Code, whereas the buzzer will reproduce exactly the signals as used in Continental Wireless telegraphy.

Furthermore, a code chart, reproducing the International Code, is engraved on a brass plate mounted between key and buzzer.

This practice set works on one or two dry cells and two outfits placed at some distance apart can be operated for sending and receiving practice.

The outfit is handsomely finished, metal parts are brass and nickel plated.

No. BEK1140 Practice Set as described.......................... **$2.50**

Shipping weight 3 lbs.

THE "ELECTRO" CODOPHONE

(Patents Pending)

This instrument imitates LOUDLY and audibly Radio Signals. It is used in learning the Morse or Continental Codes. It replaces the buzzer practice outfit, as well as the regular telegraph sounder outfit.

No. AEK1999

AMATEURS!   ATTENTION!!

The "Electro" Codophone which we present herewith is the outcome of several months of intense study and experimentation of our Mr. H. Gernsback. It supersedes our former Radiotone Codegraph, which comprised a Radiotone silent Buzzer, a loud talking telephone receiver and a key. As in all of his work Mr. Gernsback strives for simplicity. So he combined the three above mentioned instruments with one stroke into **ONE** single instrument. He combined the Radiotone Buzzer and the loud talking receiver into a single unit, not only mechanically, **but electrically as well.** This involves **an entirely new principle,** never before attempted, and on which basic patents are now pending.

What this remarkable instrument is and does.

The "Electro" Codophone is positively the only instrument made that will imitate a 500 cycle note exactly as heard in a Wireless receiver, so closely and so wonderfully clear, that Radio operators gasp in astonishment when they first hear it. And you need no receivers over the ears to hear the imitation singing spark, which sounds for all the world like a high-pitched distant powerful Radio Station. No, the loud-talking receiver equipped with a horn, talks so loud that you can hear the sound all over the room, even if there is a lot of other noise.

THAT'S NOT ALL. By lessening or tightening the receiver cap, a tone from the lowest, softest quality, up to the loudest and highest screaming sound can be had in a few seconds.

FURTHERMORE, this jack-of-all-trades marvel, can be changed instantly into our famous silent Radiotone test buzzer, simply by replacing the metal diaphragm with a felt disc, which we furnish with every instrument.

FOR INTERCOMMUNICATION. Using two dry cells for each instrument, two Codophones when connected with one wire and return ground, can be used for intercommunication between two houses one-half mile apart. Any one station can call the other, no switches, no other appliances required. No call bell either, the loud-talking phone takes care of this.

AS AN ARMY TYPE BUZZER. Last, but not least, **two Codophones** with two 75 ohm receivers can be used to converse **over miles** of fine (No. 36 B. & S. Wire), so fine that no one can see the wire. Or you can use a long metallic fence and the ground, or you can communicate **over** your 110 volt line up to several miles, using no wires, only the ground.

Full directions how to do all this furnished with each instrument.

One outfit alone replaces the old-fashioned learner's telegraph set, consisting of key and sounder, which is all right to learn the telegraph code but not the wireless codes.

The "Electro" Codophone is a handsome, well made instrument, fool proof, and built for hard work. Contacts are of hard silver $\frac{1}{8}$ inch in diameter, that will outlast the instrument. Horn and housing is of metal throughout, horn and key lever nickel plated and buffed. Three new style metal binding posts are furnished.

There is also a neat code chart and full directions enabling any intelligent young man or girl to learn the codes within 30 days, practising one-half hour a day.

No. AEK1999 The "Electro" Codophone, as described, complete.   Size: 6¾x3x2⅝". $1.50
Shipping weight 3 lbs.

THE "ELECTRO" TELEGRAPH OUTFIT.

A Complete 2 Station Telegraph Outfit for Only $1.25

THE LEARNER'S IDEAL

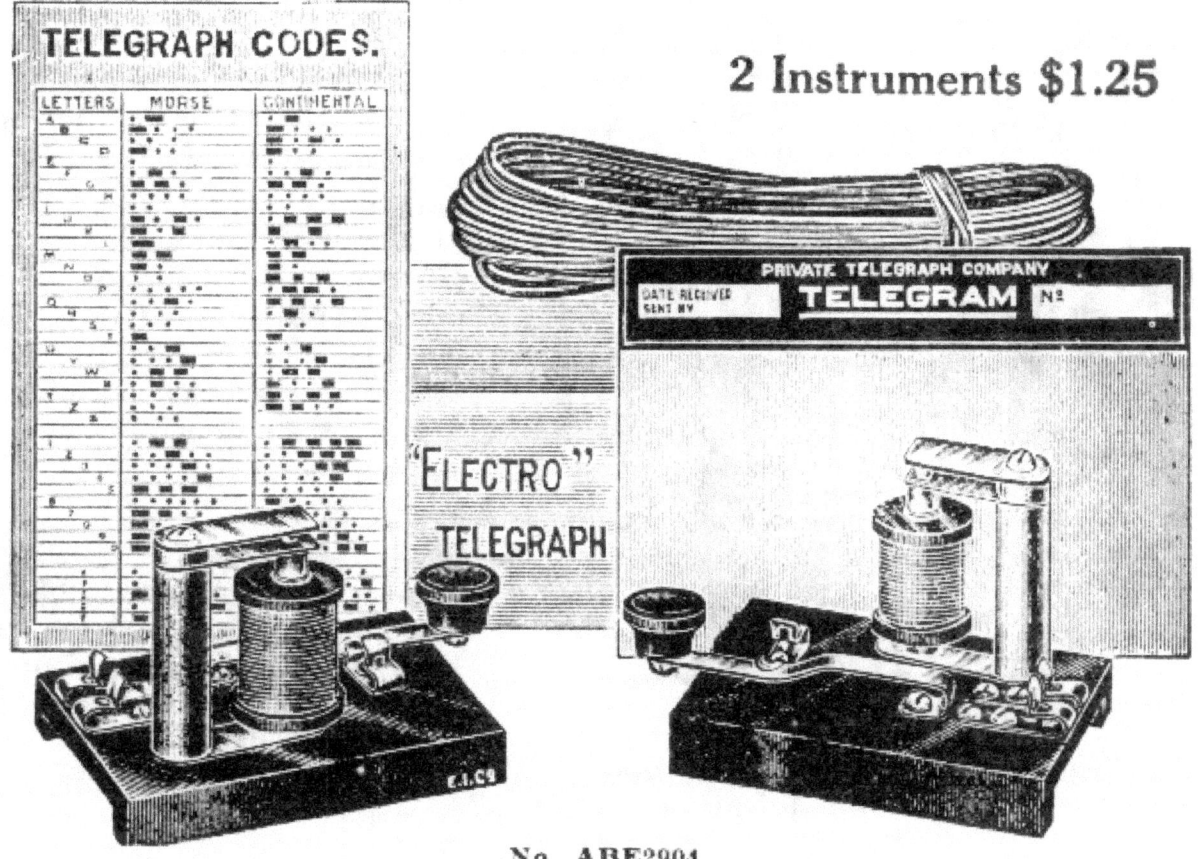

2 Instruments $1.25

No. ABE2904

While our No. BX1114 and No. BBK1115 Telegraph sets are true replicas of the telegraph instruments in use all over the world, still their cost is such as to make them somewhat prohibitive to the true learner. The "Electro" is the simplest complete telegraph outfit it is possible to make.

It consists of TWO STATIONS, for sending and receiving regular telegraph messages. Wire for connection between the two stations is also supplied. Each station consists of a key and a specially adjusted sounder which sounds just like a telegraph instrument. The entire outfit will operate on one dry battery on short lines and has been successfully used on lines one mile long. The "Electro Telegraph," while low in price, is valuable, nevertheless, for the lessons in telegraphy and electricity it can teach.

The "Electro" Telegraph does two things, and does them well, to wit:

1st—One of the instruments if used singly with one dry cell, constitutes a complete learner's set. Any normal person by devoting half an hour a day to the study of the codes, can become a telegraph operator within thirty days. By means of the code one can teach oneself telegraphy as well as if taking a course in a school. Besides the cost is ridiculously low.

2nd—If two instruments are used, between two rooms or between two houses, a complete intercommunicating telegraph system is had. Thus two persons can teach each other telegraphy in short order, and soon both can converse "over the wire" as well as any two commercial telegraph operators. It is as interesting as it is elevating; besides, telegraph operators are in great demand right now, and any young man or girl will not find it difficult in procuring a good position. Telegraph companies,—Wire and Wireless — brokerage offices, newspapers, industrial corporations and hundreds of others have always trouble in filling their open telegraph operator's positions. Here is your chance. Andrew Carnegie, Thomas A. Edison, T. N. Vail (President of the Telephone trust) were all telegraph operators once. Lots of information trickles over the wire, that boosts good operators into high positions when the time comes.

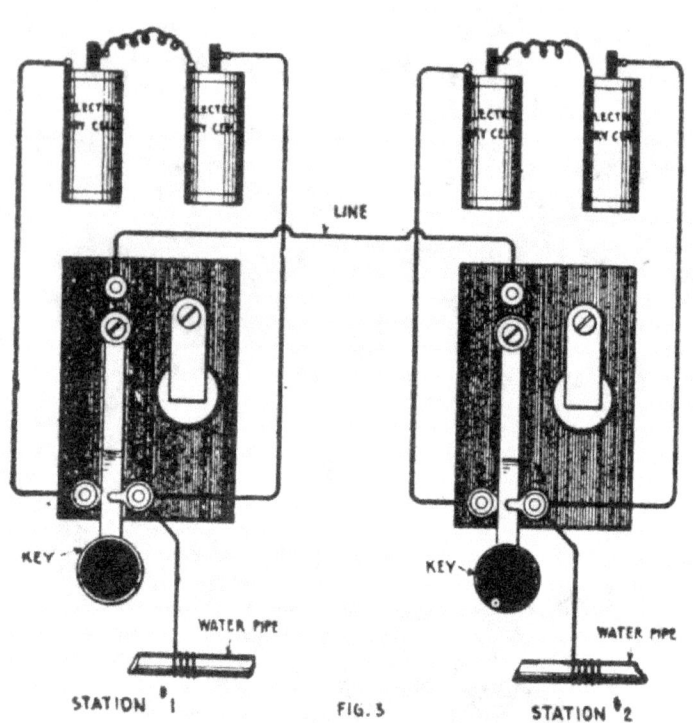

Diagram Showing Connections for Intercommunication.

It is truly remarkable how this low priced instrument has won favor. Being so simple there is naturally nothing to get out of order. As it sounds exactly like a regular telegraph sounder it is bound to teach both properly and quickly. Being made of substantial and properly put together materials it simply must last under all sorts of use, short of abuse. On account of its special connections it does not require sloppy or wasteful gravity cells but works on a convenient and low priced dry cell. If you are at all interested in telegraphy you should have this "Electro" Telegraph Outfit, the learner's ideal. Better send for one to-day; the lowest priced complete two station telegraph set on the market, barring none. Outfit is supplied in a neat box containing 2 SENDING AND RECEIVING STATIONS, WIRE FOR CONNECTION BETWEEN THE TWO, A CODE CHART, AND COMPLETE INSTRUCTIONS for direction, installation and use of the outfit. The greatest telegraph bargain ever offered to the public. Size $3\frac{1}{2}$x$2\frac{1}{2}$x$2\frac{1}{4}$ in.

This outfit can be operated by any of our dry cells.

No. ABE2409 "Electro" Telegraph Outfit, complete, as described. **$1.25**
Shipping weight 1 lb.

"The Boy's Electric Toys"

There have been other electrical experimental outfits on the market thus far, but we do not believe that there has ever been produced anything that comes anywhere near approaching the new experimental outfit which we illustrate herewith.

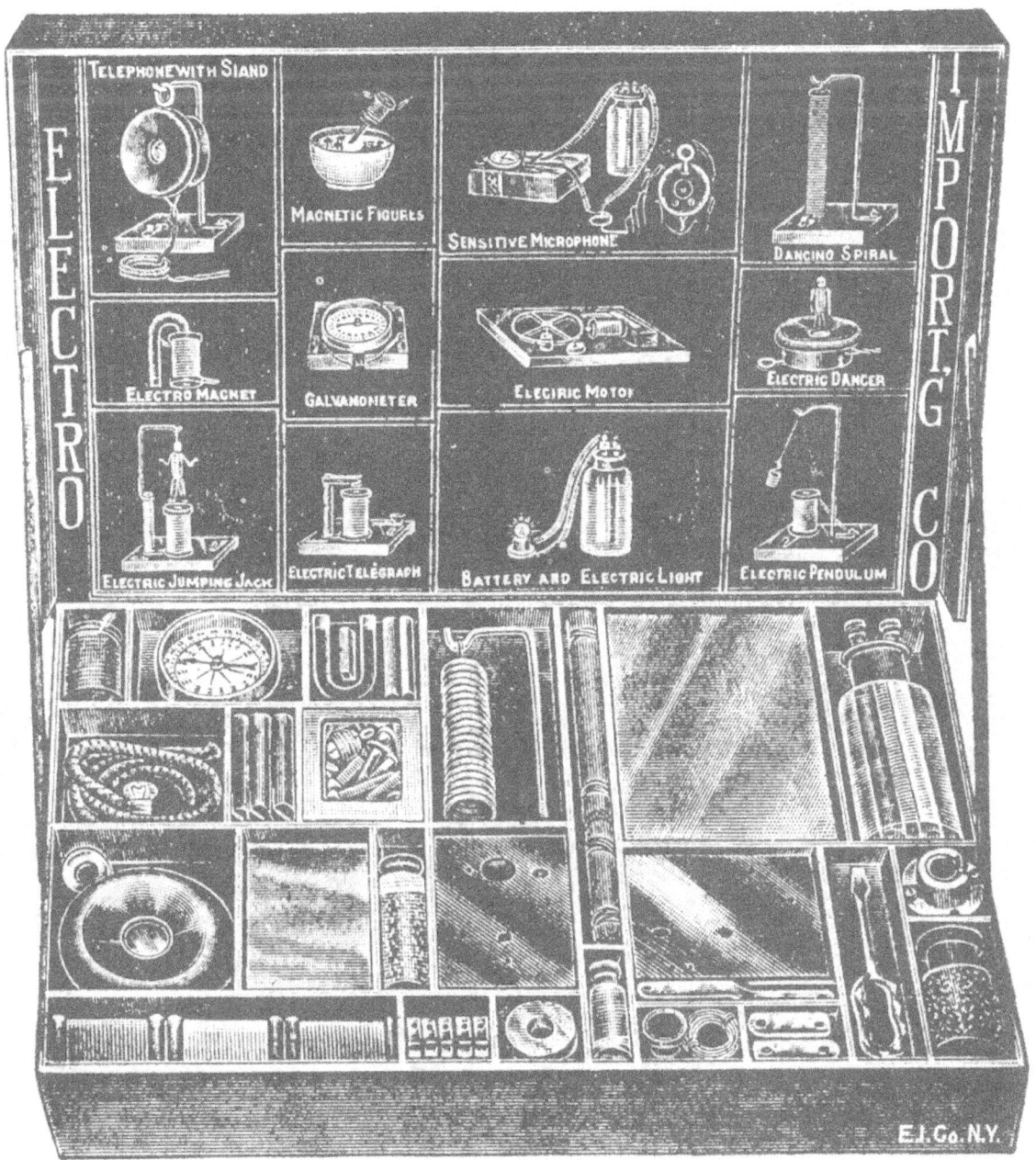

No. EX2002

"The Boy's Electric Toys" is unique in the history of electrical experimental apparatus, as in the small box which we offer enough material is contained TO MAKE AND COMPLETE OVER TWENTY-FIVE DIFFERENT ELECTRICAL APPARATUS without any other tools, except a screw driver furnished with the outfit. The box construction alone is quite novel, inasmuch as every piece fits into a special compartment thereby inducing the young experimenter to be neat and to put the things back from where he took them. The box contains the following complete instruments and apparatus which are already assembled:

Student's chromic plunge battery, compass-galvanometer, solenoid, telephone receiver, electric lamp. Enough various parts, wire, etc., are furnished to make the following apparatus:

Electromagnet, electric cannon, magnetic pictures, dancing spiral, electric hammer, galvanometer, voltmeter, hook for telephone receiver, condenser, sensitive microphone, short distance wireless telephone, test storage battery, shocking coil, complete telegraph set, electric rivetting machine, electric buzzer, dancing fishes, singing telephone, mysterious dancing man, electric jumping jack, magnetic geometric figures, rheostat, erratic pendulum, electric butterfly, thermo electric motor, visual telegraph, etc., etc.

This does not by any means exhaust the list, but a great many more apparatus can be built actually and effectually.

With the instruction book which we furnish, one hundred experiments that can be made with this outfit are listed, nearly all of these being illustrated with superb illustrations. We lay particular stress on the fact that no other materials, goods or supplies are necessary to perform any of the one hundred experiments or to make any of the 25 apparatus. Everything can be constructed and accomplished by means of this outfit, two hands, and a screw driver. Moreover this is the only outfit on the market to-day in which there is included a complete chromic acid plunge battery, with which each and everyone of the experiments can be performed. No other source of current is necessary.

Moreover, the outfit has complete wooden bases with drilled holes in their proper places, so that all you have to do, is to mount the various pieces by means of the machine screws furnished with the set.

The outfit contains 114 separate pieces of material and 24 pieces of finished articles ready to use at once.

The box alone is a masterpiece of work on account of its various ingenious compartments, wherein every piece of apparatus fits.

Among the finished material the following parts are included: Chromic salts for battery, lamp socket, bottle of mercury, core wire (two different lengths), a bottle of iron filings, three spools of wire, carbons, a quantity of machine screws, flexible cord, two wood bases, glass plate, paraffine paper, binding posts, screw driver, etc., etc. The instruction book is so clear that anyone can make the apparatus without trouble, and besides a section of the instruction book is taken up with the fundamentals of electricity to acquaint the layman with all important facts in electricity in a simple manner.

All instruments and all materials are well finished and tested before leaving the factory. We guarantee satisfaction.

We wish to emphasize the fact that anyone who goes through the various experiments will become proficient in electricity and will certainly acquire an electrical education which cannot be duplicated except by frequenting an electrical school for some months.

The size over all of the outfit is 14 x 9 x 2¾.

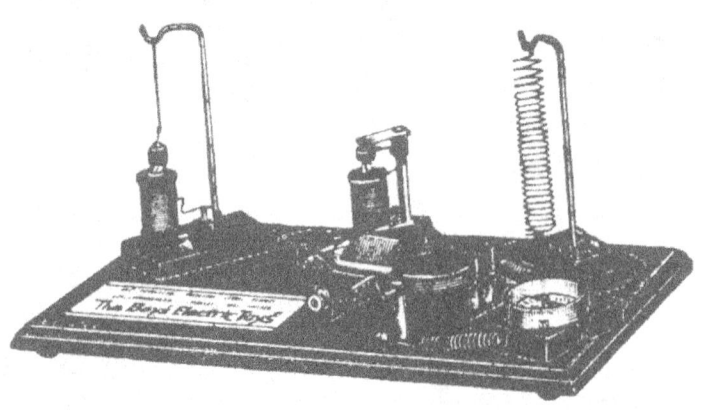

No. EX2002 "The Boy's Electric Toys," outfit as described. **$5.00**

Shipping weight 8 lbs.

Just a few things that can be made with "THE BOY'S ELECTRIC TOYS." We have not the space available to show all the other hundreds that can be made with this outfit and two hands.

Student's Chromic Plunge Battery

Here is the first low priced, as well as fool-proof chromic acid battery on the market. It is a little wonder, and for the small price we ask for it, it stands unmatched.

It is an ideal battery for electrical experimental work where a very powerful current is not required. This battery will light a 2 volt lamp for several hours on one charge; it will run a small toy motor surprisingly well; it will do small electroplating work; it is ideal for testing work; it gives a fairly steady current, **and as the zinc electrode can be pulled clear of the electrolyte, no materials are used when battery stands idle.**

Best Amalgam Zinc only is used, as well as a highly porous carbon to ensure a steadier current. We furnish enough chromic salts for 4 charges. Full directions for operation and care of battery are included. Each battery tests 2 volts and 10 amperes when set up fresh. Not over 2 amperes should be drawn from battery continuously. By using six or eight of these batteries, a great many experiments can be performed. No solution can run out of this battery if upset by accident. This makes it an **ideal portable battery,** such as for operating a bicycle lamp, or as other portable lamp, where a powerful light is not required, for boy scouts' field telegraph work, operating telegraph outfits, etc., etc. Size over all is 5 x 2 inches.

50c

No. EK999

No. EK999 Student's Chromic Plunge Battery **$0.50**
Shipping weight 1 lb.

No. 998 Carbon Rod with Binding Post, for above battery, each.. **$0.15**
Shipping weight 4 oz.

No. 997 Amalgamated Zinc Rod with Binding Post, for above, each **$0.15**
Shipping weight 4 oz.

"Electro" Solenoid

The little Solenoid electro-magnet which we present herewith is the same as that used in our "Boys' Electric Toy Outfit." It has been built especially for experimental purposes and can be relied upon in all respects. This is the only Solenoid constructed on this principle. The magnet heads are heavy fibre and the wire convolutions are wound on a brass tube which latter is rigidly attached to the fibre coil heads, making it absolutely impossible for the coil heads to pull off. This feature for experimental purposes is quite an important one. The Solenoid is wound with green magnet wire, giving it an attractive appearance. It is quite powerful, lifting 2½ to 3 pounds on two dry cells with a U-shaped core. We especially recommend

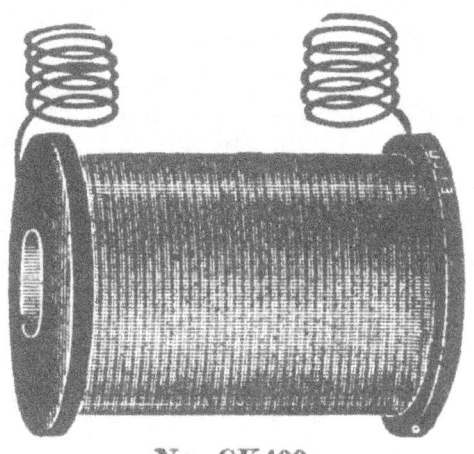

No. CK400

Type R. E. should be charged with a steady current of two amperes for ten hours. It will then give twenty ampere hours, that is a current of one ampere strong for twenty hours long, or two amperes for ten hours. Type H. O. should be charged with two amperes for five hours. Its capacity is ten A. H. It will run, for instance, a small motor which takes one ampere, ten hours at a stretch. Battery must then be charged again, after which it will give the same output hundreds of times over. We guarantee all our plates for two years if handled as per our instructions. To charge cells: Only direct current can be used. Any electrician in your town who has charge of a private plant will be glad to charge your cells. The usual rate is ten to fifteen cents per cell. If you have direct current on your premises you can charge the battery yourself. We sell charging attachments for any voltage and give explicit directions with same. When ordering state voltage of your line. If you have alternating current it will be impossible to charge storage cells unless our "Electro" Lytic rectifier No. EX12500 is used, which changes the alternating current to direct. If you have no current available, do not try to charge with dry cells. The dry cells would be spoiled inside of one hour. The charging voltage must always be higher than the combined voltage of the batteries. A fully charged storage cell, with charging current turned on has 2.5 volts. When the charging current is taken off the voltage of cell will drop at once to 2.2. This is correct. Three small Type H. O. cells will run any of our coils up to 1½ inch. Coils from 2 to 4 inch need 5 cells of type R. E.

TESTING

The only reliable way to test a storage battery is by means of a low reading voltmeter such as our No. AX4301. If the experimenter wants to know how far the battery is run down, the voltmeter readings should only be taken when the battery is actually working. Other tests are of no value whatever, as a storage battery, even if run down altogether previously, will nearly always register two volts per cell, on open circuit, as it is well known that accumulators always recuperate when standing unused. If they are put to work, however, the voltage immediately drops. Therefore, "Open" readings are of no value. The owner of a storage battery should always test it before starting in to use it; it will save him much annoyance. Each cell should be tested individually, as this is the only reliable way.

A storage battery—no matter what make—will stop working very abruptly and without any warning at all. This generally puzzles the layman a good deal, and, of course, he blames it all on the battery, as he cannot account for the drop. If he would have taken a voltmeter reading before starting in to use the battery he would have found that the cells registered about 1.85 volts each.

Never test an accumulator with an ammeter—that is, never connect the instrument directly across the battery. It is a "dead short circuit," and is not alone very harmful to the cell, but it will burn out the instrument.

No. BEK1251 Complete cell, type R. E., SEALED UP (see illustration), 20 ampere hours, 2 volts, containing 1 positive and 1 negative plate, with hard rubber connecting posts not affected by acid, separators, rubber bands, porcelain vent, directions; weight when filled with acid, 4 lbs., size of glass jar 6x6x1 in. each **$2.50**

Shipping weight 7 lbs.

No. EK1251a Glass jar for R. E. cell, 6x6x1 in................. **$0.50**

Shipping weight 3 lbs.

No. AEK1252 Complete cell, type H. O., SEALED UP (see illustration), 10 ampere hours, 2 volts, containing 1 positive and 1 negative plate, with hard rubber connecting posts not affected by acid, separators, rubber bands, porcelain vent, directions; weight of cell with acid, 3 lbs., size of glass jar 6x3x1 in., each **$1.50**

Shipping weight 5 lbs.

No. DK1252a Glass jar for H. O. cell, 6x3x1 in................. **$0.45**

Shipping weight 2 lbs.

TYPE R. E. & TYPE H. O. BATTERIES ARE SOLD WITHOUT CHARGING LIQUID

Portable Storage Batteries

ACCUMULATORS

STORAGE batteries of the portable type are coming more in favor every day. They are vastly more efficient and economical than dry cells, give an even as well as powerful current, and supply electricity cheaper than any other form of battery.

Many people make a serious mistake by considering the first cost of a storage battery as excessive, but a second's reflection will prove that it should not be considered at all. A concrete example shows this best:

Suppose electricity is required in an automobile for ignition. Six volts is the usual voltage. The cost of 6 good, dry cells to supply this current is from $1.80 to $2.40. If the automobile is used steadily such a set of dry cells will last about 30 to 40 days. This is equivalent to about 800 miles run. If any attempt is made to run lights from dry cells, their life is, of course, very much shorter. It is safe to say that the average auto owner will use from 6 to 8 sets of dry cells a year, at a cost of from $9.00 to $12.00. At the end of that period **HE HAS NOTHING TO SHOW FOR IT,** as the dead dry cells are thrown away as quickly as they are used up.

Compare a storage battery with the above case. Our No. HX555 6 volt 60 A.H. battery costs $8.00. When you receive it from us it is **FULLY CHARGED** and is good for 1,000 miles, supplying ignition only. You can, in addition to this, run from 2 to 4 headlights and a tail light from the same battery at the same time, but, of course, the charge won't last as long then. When the battery is run down, any garage in the country will recharge it from 20 cents to 25 cents. **Recharging it 8 times a year costs you, therefore, from $1.20 to $2.00.** If you recharge it yourself in connection with our No. EX12500 rectifier, the actual cost of current **IS LESS THAN 10 CENTS FOR EACH CHARGE.** Recharging it 8 times per year then costs you but from $0.60 to $0.80, **AND AT THE END OF THE YEAR YOU STILL HAVE YOUR BATTERY.**

If you require a low voltage current, where steadiness and power is a prime factor, our storage batteries will positively save you a great deal of money each year.

Our batteries have proved a wonderful success for the following purposes:

Automobile and motor boat ignition, automobile, motor boat and yacht lighting, running dental motors, physicians' cautery and medical batteries, **WIRELESS SPARK COILS,** for all kinds of experimental work, electro plating, operating window display motors, **CAMP AND TENT LIGHTING,** operating X-Ray coils, bed-room lighting, stable, barn and stair lighting, lighting children's bed-rooms, where gas, oil or candles are dangerous, **PORCH LIGHTS,** garden and lawn lighting with colored lamps for picnics and fetes, etc., **BUGGY LIGHTING,** classroom work, and a thousand other uses apparent to most anyone.

The current obtained from a storage battery is so wonderfully steady and even that you will positively never use any other form of current once you have tried our storage batteries. We are very confident of this statement.

Guaranteed "Electro" Storage Batteries

We are the first electrical mail order house in the country to offer the following broad guarantee on all of our storage batteries. **No such guarantee has ever appeared in print:**

WE GUARANTEE EACH AND EVERY ONE OF OUR STORAGE BATTERIES FOR ONE YEAR FROM DATE OF PURCHASE. WE WILL REPLACE ANY STORAGE BATTERY RETURNED TO US DURING THAT TIME, FOR A NEW ONE FREE OF ANY CHARGES. WE WILL NOT ASK ANY QUESTIONS, PROVIDING THE BATTERY REACHES US UNBROKEN, AS IT GOES WITHOUT SAYING THAT WE CANNOT CONSIDER AN EXCHANGE IF THE BATTERY IS RUINED, DUE TO AN EXCESSIVE DROP, OR HAS BEEN OTHERWISE GROSSLY ABUSED.

WE FURTHERMORE GUARANTEE TO DELIVER THE BATTERY TO YOU FULLY CHARGED AND IN FIRST CLASS CONDITION. IF IT SHOULD BE DAMAGED IN TRANSIT WE WILL BE RESPONSIBLE FOR SUCH DAMAGE AND WILL SEND YOU A NEW BATTERY, CHARGES PREPAID, PROVIDING YOU ADVISE US, AS WELL AS THE TRANSPORTATION COMPANY, WITHIN 24 HOURS AFTER RECEIPT OF THE SHIPMENT.

THE ABOVE IS THE STRONGEST AND BROADEST GUARANTEE ON STORAGE BATTERIES EVER MADE IN THIS COUNTRY.

Why this unusual guarantee? For the reason that we have more experience in building storage batteries than any other mail order house in the U. S. Our Mr. H. Gernsback—an expert in storage battery building—has been making storage batteries since 1903.

Only the best materials are used in the construction of our storage batteries, only skilled workmen are employed in turning them out. Our large factory using improved machinery and labor saving devices, makes it possible for us to turn out **GUARANTEED STORAGE BATTERIES**, at prices from 15 per cent. to 25 per cent. less than our competitors.

Only thoroughly selected and tested materials are used. The plates are made of the highest grade lead-antimony alloy. Our red lead and litharge used in filling the plates is produced in a factory that makes nothing but these two articles. The rubber jars we use have walls over 3/32 in. thick, and bottoms ¼ in. thick. **We do not use glass jars** encased in wood in our portable cells. If we did we could never guarantee safe delivery to you. Glass jars cost 50 to 60 per cent. less, but we prefer hard rubber. We use a high grade sealing compound, in which the hard rubber jars are sealed in. This compound is elastic, and acid proof, yet will not crack in the winter, nor run in the summer.

Short circuiting of our plates is out of the question as the specially treated separators which we use between the plates, make such a thing an utter impossibility.

Our carrying cases are made of treated, acid proof oak, which will last for years. **ALL OUR TERMINALS ARE OF LEAD WITH BRASS BUSHINGS INSIDE,** corrosion of terminals, therefore, the old storage battery bugaboo is entirely eliminated.

Our vent tubes, as well as the vents, are made of hard rubber and will never deteriorate. We use a patent vent, which makes spilling of acid, **even in transit,** an utter impossibility. The carrying handles are enameled and made of best Bessemer steel. They are arranged in such a manner so as to make the carrying a pleasure, not a hardship. Full directions for charging and handling our batteries are printed on an acid proof label, attached permanently to the battery.

LIFE OF OUR STORAGE BATTERIES

We are often asked how long our batteries will last under ordinary circumstances. We will try and explain:

If a battery is recharged every two months, and cared for intelligently it will be as good as new when it is five years old. We have batteries all over the country which have seen use of over 10 years and they are still doing their work faithfully. We certainly feel ourselves justified in claiming that our batteries will not deteriorate till they are over 6 years old. This is about the average life of a storage battery; our experience has proved it to be such.

SELECTING A STORAGE BATTERY

The safest way of selecting a storage battery is by knowing what the maximum amount of current is that can be drawn steadily without injuring the plates. The following will give this information:

From a 40 A.H. battery not more than 3 amperes should be drawn.
From a 60 A.H. battery not more than 5 amperes should be drawn.
From an 80 A.H. battery not more than 8 amperes should be drawn.
From a 100 A.H. battery not more than 10 amperes should be drawn.

This, of course, is for continuous work. Much larger amounts may be drawn if it is only for a few minutes at a time, but the above values will give best results in the long run.

The "Electro" Storage Batteries

Used on board of several U. S. battleships. Same style used in the Oldsmobile, Pullman and a number of other high-class automobiles for ignition and lighting.

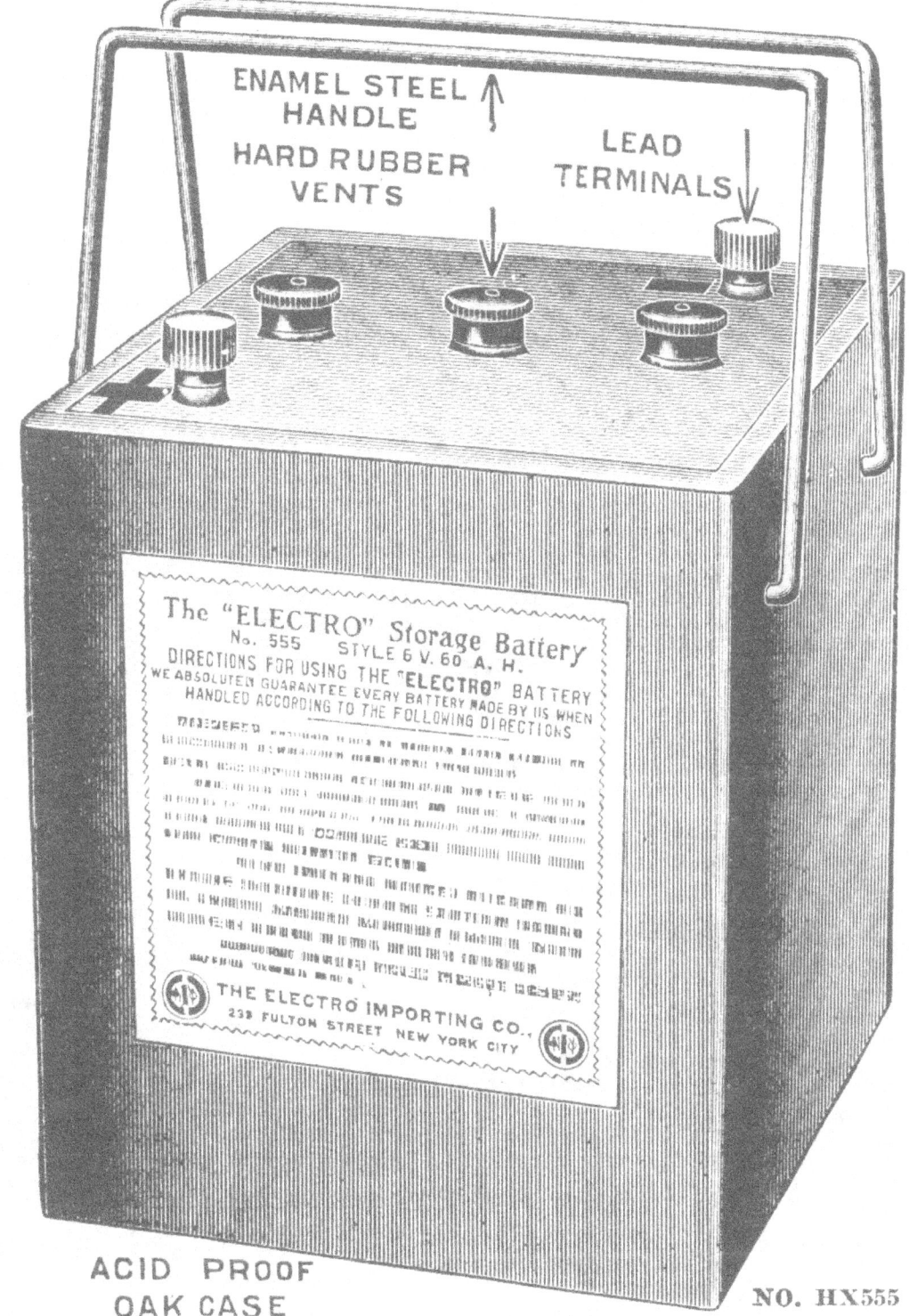

ENAMEL STEEL HANDLE

HARD RUBBER VENTS

LEAD TERMINALS

The "ELECTRO" Storage Battery
No. 555　STYLE 6 V. 60 A. H.
DIRECTIONS FOR USING THE "ELECTRO" BATTERY
WE ABSOLUTELY GUARANTEE EVERY BATTERY MADE BY US WHEN HANDLED ACCORDING TO THE FOLLOWING DIRECTIONS

THE ELECTRO IMPORTING CO.,
233 FULTON STREET　NEW YORK CITY

ACID PROOF OAK CASE

NO. HX555

The 40 A.H. size has 3 plates in each cell; the 60 A.H. size has 5 plates in each cell; the 80 A.H. size has 7 plates in each cell; the 100 A.H. size has 9 plates in each cell. **The cells are of hard rubber, NOT glass.**

This battery, which was first designed by us as an automobile battery (where it meets with the greatest abuse on account of shocks, etc.), is absolutely "fool proof." There is nothing to get out of order, and with ordinary care this battery will last for five years.

Every experimenter knows that there is practically no equal to a **good** storage battery. It is **a** pleasure to run your coil or other apparatus when you know there is lots of "juice" behind it. Our No. HX555 battery will melt a No. 10 B. & S. copper wire and gives about 200 amperes in short circuit (although we do not recommend this test as if performed often, will weaken the plates). All batteries come in treated oak cases. There is absolutely nothing to corrode. Batteries shipped fully charged, ready to use. Shipped in strong wooden, non-overturning box. Full directions on every battery.

Any garage or power plant will recharge any battery for about 25 cents. If you have direct current, you can charge it yourself at a cost of 8-10 cents by using a bank of lamps to cut down the current. If you have 110 volts A. C. this battery can be charged through a bank of lamps and our No. EX12500 rectifier shown elsewhere.

If you wish to know more about this wonderful BATTERY, send 2c. stamp and we will send you our "Treatise on Storage Batteries."

No. CIK2326 "Electro" Storage Battery, 2 volts, 60 ampere hours, size 6¼x1¾x7¼ in. Shipping weight 20 lbs........ **$4.80**

This battery is one cell only and has no oak case nor handle. It has two lead binding posts. The container is hard rubber.

No. DIE2325 "Electro" Storage Battery, 4 volts, 40 ampere hours, size 7x7x3½ in.............................. **$6.30**
Shipping weight 23 lbs.

No. FFE556 "Electro" Storage Battery, 6 volts, 40 ampere hours, size 5¼x8x8 in.............................. **$7.50**
Shipping weight 35 lbs.

No. GEK2327 "Electro" Storage Battery, 4 volts, 60 ampere hours, size 7½x7½x5 in. Shipping weight 28 lbs.... **$7.50**

No. HX555 "Electro" Storage Battery, 6 volts, 60 ampere hours, size 6¾x8x8 in. Shipping weight 41 lbs............ **$8.50**

No. ABX2328 "Electro" Storage Battery, 6 volts, 80 ampere hours, size 9¼x7x8 in. Shipping weight 60 lbs............ **$12.00**

No. AFEK2329 "Electro" Storage Battery, 6 volts, 100 ampere hours, size 12x7½x8 in. Shipping weight 68 lbs.... **$16.00**

THESE BATTERIES ARE SENT FULLY CHARGED

Hydrometers

When making up solutions for our storage batteries, electrolytic rectifiers, the Gernsback interrupter, etc., it is absolutely necessary to know the specific gravity of the acid solution.

This can only be ascertained accurately with our aydrometer. Insert the hydrometer in glass (furnished by us) and pour the acid in the glass. When reading the scale of the hydrometer a certain value is found. If the reading is too high, the solution is too strong, and water must be added until the specific gravity is correct. If too low, acid must be added, and so on. Our hydrometers are all warranted to be accurate. They come in a wooden box, with acid stand-glass, as illustrated. Size 4½x½ in. Invaluable in every laboratory.

SCALE FROM 1100 deg. to 1300 deg. **$0.30**
No. CK518 Hydrometer, as described, each.....
Shipping weight 4 oz.

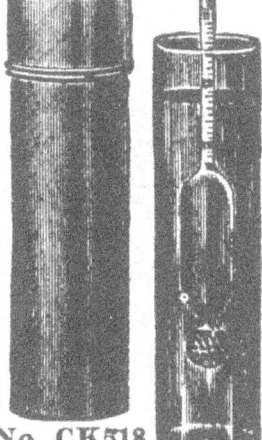
No. CK518

Dear Sirs:— Williamstown, Mass.
I received your storage battery some time ago and want to tell you **THAT IT WORKS JUST AS STATED, AND IT'S THE BEST LITTLE STORAGE BATTERY I HAVE SEEN.** I would now like you to send me your Static Machine, for which I enclose money order.

Yours truly, DOUGLAS WILD.

"Electro" Batterymeters

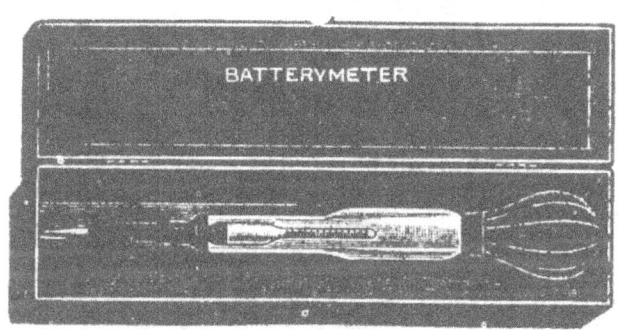

No. IE1544

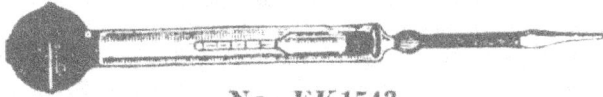

No. FK1543

Shipping weight, each style, 2 lbs.

The greatest problem to the user of storage batteries is "How fully charged or discharged is my battery?" The question of overcharge if not too great is not a serious one, but letting a battery discharge TOO MUCH is.

The batterymeter is based on the principle that the density of the electrolyte in a storage battery varies directly as the battery is charged or discharged. For instance, the density of the electrolyte is found when the battery is fully charged to be 1,250 sp. gr. Then when the reading is taken after the battery has been used, and it now shows a sp. gr. of 1,175 it means that only 40 per cent. of the full charge remains. This is readily ascertained by a glance at the table accompanying each instrument. Always bear in mind that voltmeter readings never tell the true condition of a battery. The operation of the Batterymeter is simple as it is accurate. The vent of the battery is removed, the nozzle of the meter placed in the liquid and enough liquid is drawn up by suction in the glass tube to float the hydrometer. The reading is then taken and the liquid replaced in the battery by applying pressure to the bulb and vent is replaced. The value of the Batterymeter is self-evident, and if you don't want to get caught with a completely discharged battery or seriously sulphated one, get a Batterymeter at once and see just how much juice is left in your storage battery. An invaluable instrument for the automobilist, experimenter, electrician, owner of isolated plant, etc., etc.

No. FK1543 Batterymeter, medium grade, length 10 in., hydrometer scale 15 to 35, Beaume and equivalent specific gravity. Complete, with directions in a neat cardboard box, size 12½x 2x2¼ in. Price, each **$0.60**

No. IE1544 Batterymeter. Latest type, high grade. Length 10½ in., hydrometer scale 1,150 to 1,300 specific gravity. Complete, with directions. Comes packed in a Wooden box. Very superior in every respect to the medium grade. Box size 10¼x3x3 in. Price, each **$0.95**

The "Electro" Pocket Volt and Ammeters

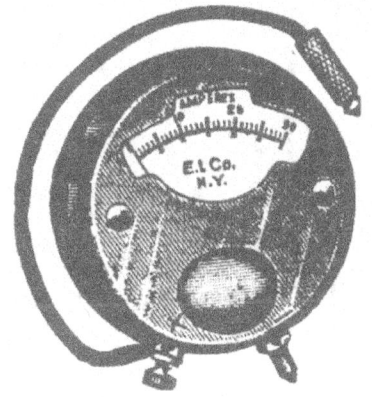

No. IE4300

These little instruments are without doubt the most compact and durable meters ever produced in this country. They are not alone extremely accurate, but they are also "dead beat," all metal construction. The finish is the finest ever seen on a domestic instrument. Permanent magnet type indicates polarity as well as current. GUARANTEED FOR 1 YEAR. It is the smallest meter made, diameter 2¼ inches, net weight 4½ oz. Flexible, detachable lead, with contact pointer furnished free. Range of voltmeters, 8 volts, ammeters, 30 amperes; combination, 8 volts, 30 amperes.

No. IE4300 "Electro" ammeter, as described, each **$0.95**

No. ABE4301 "Electro" voltmeter, as described, each **$1.25**

No. ADE4302 "Electro" Volt-ammeter, Shipping weight, any style, 1 lb. **$1.45**

The "Electro" Storage Battery Meters

This meter is so called because when properly connected to a storage battery circuit it will not only show the quantity of current passing through the circuit but also show whether the battery is charging or discharging. The normal position of the needle on this ammeter is in the centre and the needle going to the left indicates that the battery is discharging, while a deflection to the right indicates the battery is being charged.

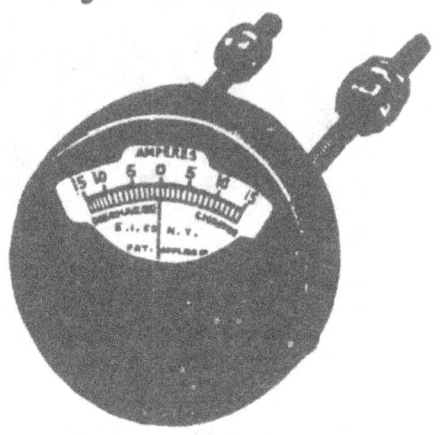

No. AIE1620

The Electro Storage Battery Meter has notified many a storage battery owner of trouble in his charging plant in time to save the battery and it will do as much for you. The meter is for direct current only 2¼ in. in diameter and ⅞ in. thick. It is made for switchboard mounting for which purpose 2 studs are provided, each having 2 nuts and washers for connections and fastening. All metal parts are brass highly nickel plated and polished. Only supplied in one scale, 15-0-15. The mechanism is of the permanent magnet type, serviceable, accurate and above all durable. This meter is particularly useful on small lighting plants such as are used on automobiles, launches, small houses, etc., etc.

No. AIE1620 Electro Storage Battery Meter, Scale 15-0-15 amperes $1.95
Shipping weight 2 lbs.

The "Electro" Volt and Ammeters

These instruments are of a lighter type than the ones mentioned elsewhere, and can be used where much space is not available. The thickness of these instruments when mounted is ¾ inch. They are especially adapted for use where they have to stand rough handling, and they will always be found to register correctly. They are used extensively in automobiles, small motor launches, movable wireless telegraph stations, etc. They have 2 solid brass studs, 1¼ in. long, with nuts and washers, leading out from back of meter, and connections can be effected very easily. The working parts are the same as the Style "A" meters. These little gauges should not be judged by their low price. calibrated dial... Finished in finest nickel plate.

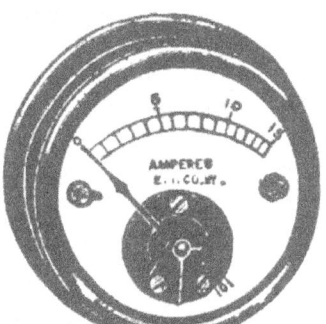

NO. CAE1041

Meters have hand

WORK ON D. C. ONLY

No. CAE1041 Voltmeter, Style B, Volts 1-12, 1¾ in. face........ **$3.15**

No. CCK1041A Ammeter, Style B, Amperes 1-15, 1¾ in. face..... **$3.30**

NOTE.—We are in an especially favorable position to quote you on any special meters you may require. Tell us what you want in the way of meters and we will quote you prices on quality instruments that will not only interest you but show you that E. I. Co. prices are always the lowest and delivery the best. Send for a quotation to-day.

A complete chapter on "METERS AND MEASURING INSTRUMENTS" is contained in the "EXPERIMENTAL ELECTRICITY COURSE" in 20 lessons, which is given FREE with one year's subscription to the "Electrical Experimenter Magazine." See announcement on back cover of catalog.

The "Electro" Magnets

(Patented Dec. 20, '10.)

For Sounders, Bells, Relays and Wireless Instruments, etc.   Illustration full size.

Our "Electro" Magnets are well made with fibre ends.   Copper wire used being double cotton covered. Magnets are covered with mottled paper.   The core is of very best soft Norwegian iron, and each core at the top has a silver contact riveted into same as shown, which may be used for making contact. This is an important feature patented by us and serves to make a contact which of course eliminates another contact arm or standard.

Size over all 15/16 by 1½; diameter of core ⅜ inches.   Three styles carried in stock.

No. BKO1107 "Electro" Magnet, 20 ohms **$0.20**
No. BEO1108 "Electro" Magnet, 50 ohms **$0.25**
No. CKO1109 "Electro" Magnet, 75 ohms **$0.30**

Shipping weight, each style, 2 oz.

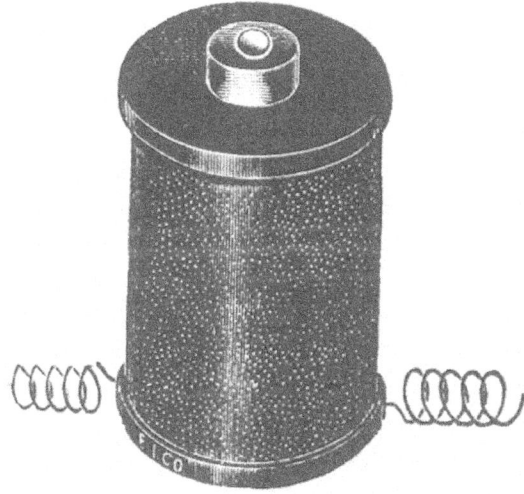

No. BKO1107

Carbon Grains and Balls for Telephone Transmitters

These grains and balls are of the best grade Carbon, the same as used in telephones.

No. AE6085   One bottle carbon grains **$0.15**

Shipping weight 4 oz.

No. BE6075   One bottle granulated carbon balls.............. **$0.25**

Shipping weight 4 oz.

Zinc Spark Ball

For wireless high frequency and other experiments. 1¾ in. diameter, 1 in. high, with split brass tube ⅛ in. diameter, highly polished.

No. BE6018   Zinc Spark Ball, each.   **$0.25**
Shipping weight 4 oz.

No. BE6018

No. AE6085

The "Electro" Dry Battery

This is one of the best batteries ever manufactured and we do not believe a better one can possibly be manufactured for the price.   This battery, which is in extensive use in the trade, sells entirely upon its merits.   It is a well-known fact that most dry batteries when standing on the shelf unused for a short period quickly degenerate and lose most of their power, as the local action is too strong; the "Electro," however, can stand idle for a considerable length of time without disintegrating, as there is absolutely no local action when battery is not in actual use.

A very important point about this battery is that it can be partially recharged quite easily when once run down, and we guarantee that it will come up to nearly its original standard when thus fully recharged.   Any one can do it. This battery gives 1½ volts and from 18 to 20 amperes.   This information will only be given with battery purchase.

No. DK1001   "Electro" Dry Battery.   Size 2½x6 in. **$0.40**
Shipping weight 3 lbs.

NO. DK1001

The "Electro"-Lytic Rectifier

This article fills a demand, which of late has become burning. For the past years we have been flooded with mail to supply a rectifier to change alternating current to direct. As is well known, alternating current **can not be used to charge storage batteries** and for a great many other purposes also, direct current only can be used. This especially is the case with spark coils for wireless, etc. On the other hand, the experimenter who has an alternating current supply, when using our rectifier can perform many experiments otherwise impossible.

NO. EX12500

Our rectifier works on any cycle alternating current up to 110 volts. It must be used in connection with a lamp or water resistance and cannot be connected to the current supply without the resistance in series with it. The efficiency of the 4 jar rectifier is 85%, AND THERE IS NO LEAKAGE such as is usually found in other rectifiers. The 4 jar rectifier furthermore USES BOTH SIDES OF THE CYCLE, which accounts for the high efficiency. Of course, a one or two jar outfit may be used successfully, and EVEN ONE JAR ALONE rectifies alternating current to direct but the efficiency is necessarily low, as only one half of each cycle is used.

However, for experiments using little current, the one jar type is very satisfactory. The 4 jar type passes as much as 5 amperes and can be used continuously (as for instance charging storage cells) with 2½ amperes. The one and two jar types pass 1/3 of the above amperage. Our rectifiers come ready for use. All you have to do is to dissolve the salts in HOT water and fill in jars, and the rectifier is ready. The 4 jar type comes with a wooden tray (see illustration) which holds the jars. No tray is furnished if jars are bought separately. The covers are of heavy porcelain with polarity marks in plain sight. Binding posts are hard rubber, impossible to short-circuit or to shock you. The lead and aluminum plates are very substantial and the latter will last for months. If used up they can be renewed at small cost and replaced with new ones in a few minutes.

New aluminum plates will **not** be sold separately except to users of the rectifier. When ordering renewal plates it is ABSOLUTELY REQUIRED that you give us your order number or date of purchase.

OTHER PARTS ARE NOT SOLD. Directions and diagrams **only** furnished with rectifier. Every rectifier is guaranteed.

No. EX12500 Four jar rectifier, as described, with tray........ Size 8½x10½x10½. Shipping weight 30 lbs.	**$5.00**
No. ABE12501 One jar rectifier, as described................... Size 8x5½. Shipping weight 5 lbs.	**$1.25**
No. BK12502 Renewal aluminum or lead plates, each........... Shipping weight 1 lb.	**$0.20**
No. BE12503 Renewal salts (charge for one jar)............... Shipping weight 1 lb.	**$0.25**
No. DK12504 Glass jar for rectifier, size 4½x6½ in............. Shipping weight 4 lbs.	**$0.40**

Dear Sirs:— Portland, Ore.

I received my Rectifier in very good condition and I am pleased to tell you that it is working fine. WALTER OLSON.

The "Electro" Rheostat-Regulator

PORCELAIN BASE

(PATENTED FEB. 1, 1910.)

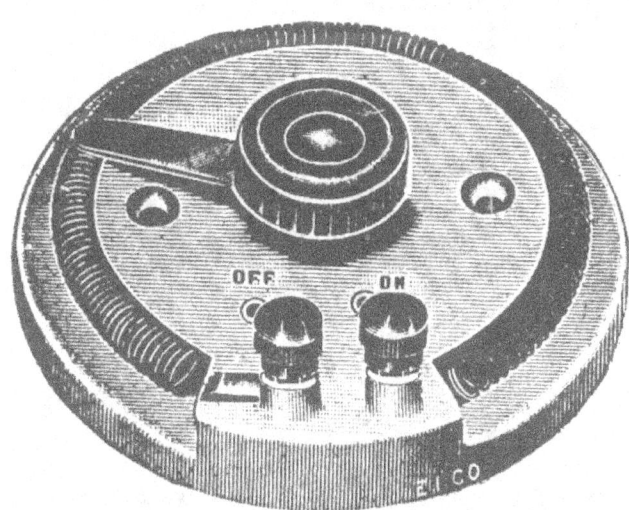

NO. FK5000

This little current regulator makes a valuable addition to any wireless set where it is used to regulate the battery current.

With battery lamps it is very valuable, where it is used to prevent the lamps from burning out on account of too strong a current, etc.

In connection with small motors it will regulate the speed more accurately and more gradually than could be done by any other means. This feature makes it very desirable for Dentists, Doctors, and all those who need an effective regulator. In connection with cautery work it is indispensable, as any degree of heat can be obtained,—due to the very fine regulation.

One of its real values is in providing a small field regulator for small motors, and dynamos giving in that way a voltage and speed regulation not obtainable with the usual type of expensive step type rheostats.

It also makes an excellent automobile lamp dimmer where it can be used to cut down the glare yet not dim the lamp enough to cut down the light too much. This is a fault of many of the present auto dimmers now in use.

Advantages over other small rheostats: Gradual and accurate regulation of current; great current capacity; little heating, resistance coil aircooled; no concealed parts; impossible to get out of order. **PORCELAIN BASE. CANNOT BURN NOR CHAR.**

For electro-plating work it will be found indispensable. A gradual increase of current is especially necessary for fine work.

The wire used in this regulator is the finest imported high resistance wire. It will positively not rust, break nor bend, even under a constant load of 3 amperes. This we guarantee in every instance. The groove which holds the spiral is ( ) shaped (PATENTED), which makes it impossible for the coil to fall out or become dislocated once wound in place. Large hard rubber handle (1 in. diameter) is provided, allowing rapid and smooth turning of switch blade. **Don't use a rheostat with the resistance wound on fibre that will smoke and smell as soon as current is turned on.**

Resistance is 10 ohms. Maximum capacity, 3 amperes continually; size, 4 inches diameter, thickness of base 13/16 inch.

No. FK5000 Rheostat-Regulator (patented). Price............... **$0.60**

Shipping weight 2 lbs.

A complete chapter on "RECTIFIERS" is contained in the "EXPERIMENTAL ELECTRICITY COURSE" in 20 lessons which is given FREE with one year's subscription to the "Electrical Experimenter Magazine." See announcement on back cover of catalog.

The "Electro" Tesla Transformer

(Patent Applied for)

We hardly need mention that the Tesla Transformer is one of the most marvelous pieces of apparatus ever invented and there are thousands of experiments and demonstrations that can be performed with this apparatus.

We had this transformer under consideration for years before we built it and we had to build dozens of different models before a perfect apparatus was produced. We do not hesitate to say that for the price this apparatus is the greatest bargain ever offered by us. Although it is the lowest priced Tesla Transformer in this country it does not mean that it is less efficient than one costing $50, but as we are manufacturing the transformer in very large quantities it is possible to greatly reduce the cost. The manufacturer who only builds one or two at a time must ask a great deal more than we, making two or three hundred at a time.

NO. EEK7000

The construction of this transformer has been simplified to such an extent that it is not only absolutely "fool proof" but we guarantee that it will do anything and everything any standard Tesla transformer ever did, or will do. The primary is wound with the best imported Pirelli cable and the secondary with the best DOUBLE INSULATED enamel wire. The secondary is insulated by solid hard rubber brackets as shown in cut. It is also provided with two well finished coil ends. The insulation is absolutely perfect, the base is heavy quartered oak. Our usual hard rubber binding posts are attached as shown.

The Tesla Transformer is an apparatus which steps up the frequency of the secondary of a spark coil, to many hundred times its original value. It is this incredibly high frequency that produces the wonderful phenomena you have seen performed on the stage by some clever electrician, all of which can be duplicated with our transformer.

The "Electro" Tesla Transformer can be operated in conjunction with the following apparatus; it is of no value without them:

First, a spark coil or transformer coil; second, a set of condensers; third, a spark gap. One of our 1 in. spark coils is sufficient to operate the transformer, but we would always recommend getting a 2 in. coil with which the best results are to be had. Our ½ K.W. transformer coil in connection with our Electrolytic interrupter will give still more wonderful effects. Two of our 1½ pint Leyden jars No. ABE9222 or any other sending condenser of the RIGHT CAPACITY can be used in connection with this transformer. A single spark gap completes the entire outfit. IT MUST BE UNDERSTOOD THAT THE TESLA TRANSFORMER CANNOT BE USED WITHOUT THE ABOVE MENTIONED ACCESSORIES.

To get perfect results and to obtain the best possible effects we would greatly recommend the following outfit:

No. EEK7000 "Electro" Tesla Transformer. Shipping weight 11 lbs. **$5.50**
No. IFK1089 Spark Coil, 2 in. Shipping weight 8 lbs. **9.60**
No. ABE9222 2 Leyden Jars. Shipping weight 6 lbs. **2.50**
No. GE9220 Spark Gap. Shipping weight 1 lb. **0.75**

No. AHCE7001 Outfit complete. Shipping weight 25 lbs. Total **$18.35**

OPERATION

The connections are made as shown in diagram. It is of highest importance that the connecting wires from the secondary of the spark coil to the Tesla Transformer and to the condensers, are heavy (not less than No. 14 B. & S.). There should be as little wiring as possible, no loops or coils, and if possible all wires should be of equal length. When connections are made

WIRING DIAGRAM

The "Electro" Tesla Transformer (Continued)

as shown, start operating. First test for spark length of the Tesla, by attaching two stiff copper or brass wires to the secondary coil leads (see illustration No. EEK7000). Leave a gap of about 2 inches between the two wires. Then start the coil. If the Tesla spark is not long enough, the spark gap must be adjusted, until best results are obtained. It is important that the zinc spark plugs have POINTED ENDS. Plugs with flat ends cannot be used. A few minutes experimenting will give the right Tesla spark. It gives a crackling noise, which becomes louder as the spark of the zinc spark gap is lengthened.

All Tesla experiments should be performed in absolute darkness, as then the best results are obtained. When everything is adjusted and working right, the wires leading from the secondary (Fig. EEK7000) give a blue brush. This works best when there is no spark discharge between the wires.

Next take off all wires from the Tesla Secondary, which we will call T. S. hereafter. Operate apparatus and both coil ends of the T. S. will show the "Fire Wheel." The brush composing the wheels actually turns in unison with the interruptions of the spark coil. Next take a piece of metal (anything will do), and, holding it in your hand, approach one of the T. S. You will observe a large brush as you approach, or a spark from 2 to 3 inches long will jump into the metal piece without you experiencing the slightest sensation.

Although the voltage of such discharges runs into the hundred thousands, it is harmless. A spark can also be drawn with the bare hand without harm, although it stings a little. Touch one of the T. S. with a piece of metal which you hold in your hand. With the other hand approach slowly and carefully the other T. S. About 4 or 5 inches away from it, a large beautiful brush will be drawn from the hand, very weird in character. Geissler and Tesla tubes light up when brought near the Tesla Transformer, without actually touching any part of it. They are usually operated by holding one end of the tube in the hand and drawing a spark into the other end of the tube by approaching one of the T. S. Ground one of the T. S. and attach a piece of wire to the other T. S. An enormous brush will be observed on the free wire. Another interesting experiment is performed by running two fairly stiff copper wires parallel to each other and about 2-4 inches apart (Fig. 1). When the frequency is high enough the space between the wires will be filled with light, while the ends will show a flame-like discharge. Small flames also play continuously on both sides of the wires.

Two rings formed of copper wire, the small ring placed into the larger one, and both connected to the poles of the transformer, will show a very pretty discharge (Fig. 2). A similar experiment is performed by forming two stiff wire loops, of heavy copper wire. The two loops are placed one into the other, and both are in the same plane. The space between the two loops will be filled with light when the coil is in operation (Fig. 3).

In Fig. 4, R. R. represents two thin hard rubber sheets. A thin silk covered wire, No. 36, B. & S., is glued in form of a name on the face of each sheet. The back of the sheets is covered with a piece

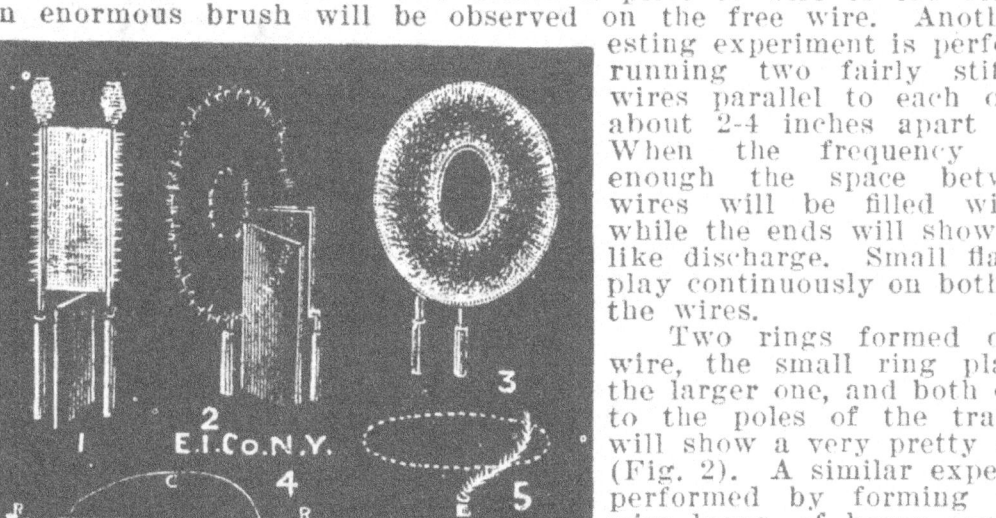

TESLA EXPERIMENTS

of tinfoil. Both are connected by a thin wire, C. The terminals of the transformer lead to the name or characters as indicated in the illustration. After a little experimenting the point will be found where the name or characters will stand out brilliantly.

Fig. 5 shows another experiment to be produced in the dark. A very thin, bare copper wire attached to one end of the transformer, rotates in a circle. The length of the bare wire must be ascertained by experiment.

A short piece of thin, cotton covered copper wire when attached to one end of the transformer is enveloped in a beautiful light discharge. (Fig. 6.)

These coils can be used to advantage for all kinds of work where high tension currents are required, such as wireless telegraphy, lighting Geissler and X-ray tubes, and for other interesting experiments,

Size of base 16 in. by 7 in. Height over all 6½ in. Shipping weight 11 lbs.

No. EEK7000 "Electro" Tesla Transformer, as described....... **$5.50**

The "Electro" Nickel Plating Outfit

We pride ourselves to be the first concern in this country to offer for sale a small nickel plating outfit within the reach of everybody.

Please do not take this outfit for a toy, as it is able to do all kinds of small work just as thoroughly as any nickel-plater could do it. It does not differ in any respect from commercial outfits, except in size. If you live in a small town you can earn quite a little during your spare time by nickelplating small objects for which work there is always a ready demand. Our outfit for neatness and efficiency is unmatched. It has two heavy nickel Anodes, heavy suspension frame resting on the glass tank and three connection rods. The object or objects to be plated are hung by means of copper wires, on the central rod, and as there are two Anodes, **a uniformly fine deposit is obtained** that cannot be matched. Good plating can be done within 15 minutes, heavier plate from 20 to 30 minutes. The Anodes are very substantial in size and will last for months.

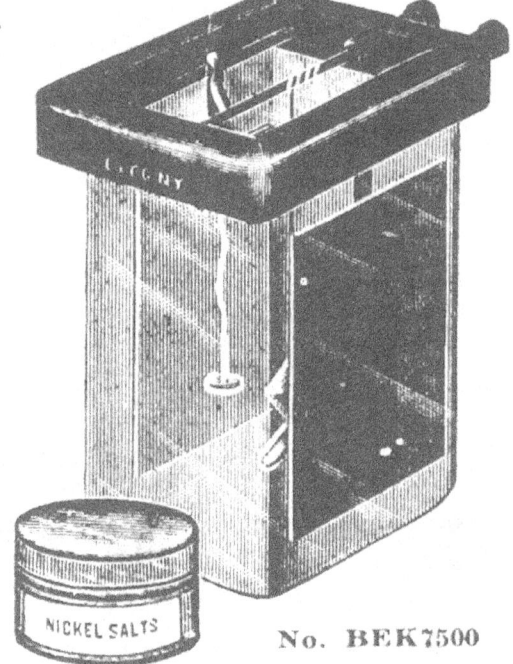

No. BEK7500

The size of the tank is 3¾x4x6 in., and the largest object that can be plated with this outfit should not measure over 3¼x2½x5 in.

To operate the outfit we recommend 3-4 of our No. BGE10050 Gordon Batteries. With these the outfit can be worked for a long period. Dry cells cannot be used under any circumstances. Any battery giving from 3-6 volts and from 5-10 amperes STEADY CURRENT can be used.

If you do not care to polish the work by hand we recommend our utility motor No. ACX325. You will then have a complete commercial outfit and you can turn out work rapidly.

Our outfit is greatly recommended to **jewelers, schools, repair shops, garages, electricians** and all those desiring to do nickel plating economically.

Complete printed instructions come with each set.

No. BEK7500 "Electro" Nickel Plating Outfit (without batteries) complete **$2.50**
Shipping weight 4 lbs.

No. GK7501 Extra Anodes, each **$0.70**
Shipping weight 4 oz.

No. EK7502 Extra Bottle Nickel Salts, each.................... **$0.50**
Shipping weight 1 lb.

No. DK7503 Extra Glass Tank, size 3¾x4x5 in., each........... **$0.40**
Shipping weight 3 lbs.

The "Electro" Galvanometer

NO. DK1376

F ROM the time that galvanic electricity was first found to possess the power of deflecting a magnetic needle to the present time the galvanometer has been used for the detection and measurement of minute and even large quantities of electric current, as a voltmeter, etc. The "Electro" galvanometer very adequately fills the bill of the experimenter who requires a cheap but good current detector as well as a polarity indicator, etc. The number of experiments in which the galvanometer can be used are almost countless and when the low price is considered we are sure you will not care to be without so valuable an article. Galvanometer has a compass 1¾ inches in diameter and is mounted on a polished hard wood base 3¼x3¼ inches in size with binding posts for connections. A real useful bargain.

No. DK1376 The "Electro" Galvanometer as described............ **$0.40**
Shipping weight 4 oz.

The "Electro" Lighting Plants

Adapted to launches, automobiles, private houses, moving **picture** theatres, shows and bungalows.

We have been literally flooded with inquiries for prices on complete electric lighting plants for various purposes, and so have finally developed a new and improved line embodying the most up-to-date principles and eliminating the weak point of all others.

These lighting plants are used in connection with our marvelously efficient Tungsten filament lamps, which give 1 C.P. for every watt delivered by the dynamo. We have spared no expense in getting together the very best of reliable apparatus for this work: The dynamo is similar to that used by the U. S. Government, as is also our special air-cooled 2 cycle gasoline engine.

The storage battery is our well known "Electro" 6 volt, 60 ampere hour size, and on the larger size plants any number of them can be connected in series or multiple to give the desired output. Our automatic cut-out of the magnetic type has been developed for automatically preventing the battery from discharging back through the dynamo, if the latter's speed and consequently its voltage, falls below that of the battery at any time.

Electric lights are the safest and most convenient illuminant known to man, at the present time. Their cost is lower, all things considered, than any other forms of illumination; oil lamps are extremely dangerous and moreover very unhealthy. Do you know that one gas light consumes as much oxygen in one hour from the air in a room as six full grown people? Do you know that more fires are caused annually by the careless use of matches, oil lamps, etc., than all other causes put together? This is a fact proved by Insurance and Fire statistics.

In figuring on lighting outfits, the following rules will be of service. The watts are obtained by multiplying the volts by the amperes. Conversely, the volts or the amperes, are found by dividing the watts by the known quantity. About 2.6 volts dynamo voltage should be allowed for each storage cell to be charged in series. On dynamo outfits without batteries, the lamp voltage should correspond to that of the dynamo. On battery and dynamo outfits combined, the lamp voltage should be the same as that of the battery.

We have arranged for lighting outfits in various combinations of equipment, so that practically all ordinary requirements are covered. The individual apparatus is also listed separately, so that the engine, dynamo, etc., may be purchased as desired by the customer.

In the following description the most important apparatus will be explained:

SWITCHBOARD

Our Lighting Plant Switchboard here shown is made up of an insulating panel, upon which are mounted the various instruments and controlling devices for operating our lighting sets. A voltmeter and ammeter is supplied for reading the dynamo potential, and the charging as well as the discharging current from the storage battery. A No. FK5000 Rheostat is furnished and connected so that the field current strength of the dynamo may be adjusted and thus control the voltage of same. An automatic, magnetic cut-out serves the purpose of disconnecting the dynamo circuit from the storage battery, whenever the former's voltage falls below the normal charging value. Fuses and switches of ample capacity are fitted to the panel, for both the dynamo main circuit and the lamp circuit. The voltmeter circuit has a switch in it, so that it need not be left in circuit, except when reading the voltage. The ammeter is always in circuit. The board is completely wired up, with all connections made, and all the purchaser has to do is to connect up a few wires from the dynamo and storage battery to the marked terminals on the panel. Blue-prints are furnished showing clearly these simple connections so that anyone

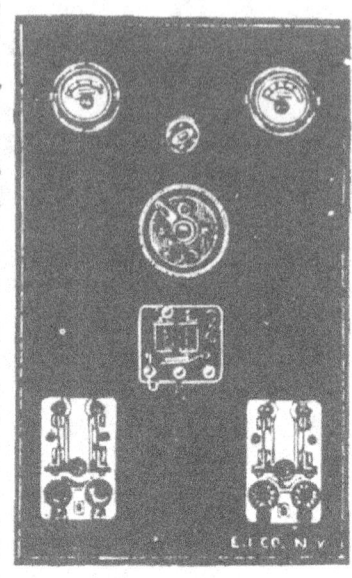

No. AFX4640

can make them in a few minutes. The size of this switchboard is 24 inches by 15 inches. It is simple and substantial, giving a neat and ornamental appearance to any station.

No. AFX4640 Switchboard containing 1 No. DEK4630 automatic cut-out; 1 No. CX1037 voltmeter; 1 No. CX1039 ammeter (Style A); 1 No. FK5000 field rheostat; 1 Voltmeter switch; 2 switches and fuses. Size 24x15 inches...................................... **$16.00**

Shipping weight 20 lbs.

CUT-OUT

AUTOMATIC CHARGING CUT-OUT

Our automatic charging cut-out is absolutely necessary whenever a dynamo is to charge a storage battery. It serves to automatically disconnect the dynamo circuit from the battery, whenever the former's voltage falls below that of the battery. If this is not done, the battery will discharge itself through the dynamo probably reversing its polarity, and it may badly sulphate or over-discharge the battery. This cut-out has two windings on it, a fine wire shunt winding and a heavy wire Series Coil. It is well made in every particular and is rugged enough for automobile service, etc., etc. It closes itself as soon as the dynamo reaches the proper speed and voltage, which are interdependent and proportional. Diagrams furnished with every instrument.

No. DIE4630

No. DIE4630 Automatic magnetic cut-out (mounted on composition switchboard) for preventing storage battery discharging back through dynamo **$4.95**

Shipping weight 2 lbs.   Size 2¾x2½ in.

Gentlemen:—   Saratoga Springs, N. Y.

I have just received your reply concerning the wire I wound on that Magneto generator and was exceedingly pleased to note your kindness in such a thing as this. I bought considerable goods from you and they were always satisfactory in both quality and price.

CHAS. M. COGAN.

Gasoline Engines

The gasoline engine is a newly developed air cooled type, which has been adopted by the U. S. Government for many purposes, including the driving of wireless generators. It is well made in every particular and develops ½ H. P. easily, with speeds variable from 800 to 2000 R.P.M. Its net weight is 23 lbs. and it measures 12½ in. high, by 1 sq. ft. floor space. Cylinder has 1¾ in. bore by 1¾ in. stroke. The flywheel has a diameter of 8 in., the pulley a diameter of 1¾ in. It is air cooled by a fan built on the flywheel, 2 cycle type, jump spark ignition. Fuel tank in base.

No. CHX4600 ½ H.P.

No. CHX4600 ½ H.P. Gasoline Engine, complete with all ignition apparatus and muffler **$38.00**

Shipping weight 60 lbs.

150 Watt Dynamo

Our 150 watt, commercial type, heavy duty, 8 volt, 19 ampere, D. C. shunt wound generator, is of the latest improved type with ample, well lubricated bearings. It has a liberal size commutator, of drop forged segments. Special high conducting brushes are supplied, and this is very important for a low voltage dynamo. The frame is very heavy and substantial and thus the machine can be run continuously without undue heating. A solid steel shaft passes through the armature and has a steel driving pulley secured firmly to it. The armature is built up

NO. CEEK4620

of thin annealed silicon steel discs, properly slotted to accommodate the windings. The windings are thoroughly impregnated with electrical varnish and baked. This machine is not a toy, but built for industrial purposes. It may be run 20 hours a day.

To deliver its proper output it should be run at 2200 R.P.M.

No. CEEK4620 Special 150 watt, 8 volt, 19 ampere dynamo.
Price, complete with pulley **$35.50**

(Shipping weight 45 lbs. Size 9x10x12 in.)

Complete Plants

No. IEX16000 19 LIGHT, 150 WATT, LIGHTING OUTFIT

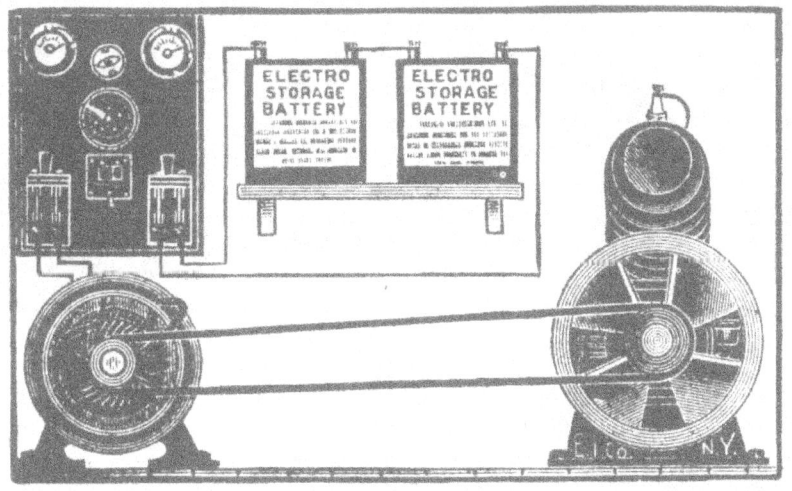

This outfit is the most flexible and widely adaptable lighting plant we list. It forms an ideal plant for motor boats, automobiles, bungalows,, camps, etc. The capacity of 19 lights is for the dynamo and engine, considering 8 C.P. 8 volt Tungsten lamps being used. The outfit with 1 No. HX555 storage battery can light 6 to 7, 6 volt, 6 C.P. Tungsten lamps for from 7 to 8 hours. The dynamo can easily recharge 2 No. HX555 60 A. H.

No. AKEX16003

batteries at once, and these 2 batteries will give twice the above C.P. output on discharge. The outfit consists of above No. CEEK4620 150 watt dynamo, No. BGX4600 ½ H.P. gasoline engine, No. DEK4630 automatic cut-out, and No. HX555, 6 volt, 60 ampere hour storage battery, capable of lighting 6 to 7, 6 volt, 6 C.P. lamps or their equivalent for 7 or 8 hours at each charge; 12 ft. 1 in. leather belting.

No. IEX16000 Complete outfit, as described...................... **$95.00**

Shipping weight 130 lbs.

No. AKEX16001 COMPLETE ELECTRIC LIGHTING OUTFIT

The same plant but with 2 No. HX555 batteries, giving twice the above candle power 7 to 8 hours, or the same C.P. for twice the above time in hours.

No. AKEX16001 Outfit complete **$103.50**

Shipping weight 170 lbs.

No. EAX16002 COMPLETE ELECTRIC LIGHTING OUTFIT

For those having motive power at hand, such as in motor boats, automobiles and the like the following outfit will be sufficient and very adaptable. Outfit consisting of above No. CEEK4620 generator, No. DIE4630 automatic cut-out, one No. HEK555 storage battery, and 12 ft. 1 in. leather belting.

No. EAX16002 Complete outfit, as described. Price............ **$51.00**

Shipping weight 85 lbs.

No. AKEX16003 COMPLETE ELECTRIC LIGHTING OUTFIT

This is our best outfit. It is the best that we build, and as good as money can buy. We guarantee that this outfit will give satisfaction in all respects whatsoever. Capacity 19 lights, 150 watts.

This outfit consists of

1 No. CEEK4620 8 volt, 19 ampere, 150 watts D. C. dynamo, 2200 R. P. M., shunt wound $35.50

1 No. AFX4640 Switchboard containing 1 No. DEK4630 automatic cut-out; 1 No. CX1037 voltmeter, 1 No. CX1039 ammeter (Style A); 1 No. FK5000 field rheostat; 1 voltmeter switch; 2 switches and fuses. Size 24x15 inches $16.00

1 No. CHX4600 ½ H.P. air cooled gasoline engine, complete..........$38.00

1 No. HEK555 Storage battery $8.50

4 Extra fuses $0.20

60 ft. No. 10 R. C. S. B. copper wire (making 4 mains 15 ft. long for generator and battery lines) $2.45

15 ft. No. 14 R. C. S. B. copper wire for field rheostat line........... $0.45

No. CX4650 12 ft. 1 in. leather belt............................. $3.00

1 No. BE517 Hydrometer $0.25

1 Set plants for installing plant

No. AKEX16003 Complete Electric Lighting Outfit, as described **$105.00**

Shipping weight 185 lbs.

A complete chapter on "ELECTRIC LIGHTING PLANTS" is contained in the "EXPERIMENTAL ELECTRICITY COURSE" in 20 lessons which is given FREE with one year's subscription to the **Electrical Experimenter Magazine.**" See announcement on back cover of catalog.

The "Electro 8-10" Dynamo

No. AEX810

When we come out flat-footed and say that this machine is the very finest low voltage dynamo in this country, we make the statement without reserve. No "although," no "ifs," no "buts." The "Electro 8-10" is a marvel all through. It is built like a watch, and solid as a gun. For workmanship, efficiency and lasting qualities it stands on the very peak of excellence. But words and praise alone won't do—you must see the machine, you must have "lifted" it and you must have seen it run, to appreciate it. We say seen, because it works so marvelously easy, that it can hardly be heard when it runs at full speed, under full load. Wherever there is surplus power, be it gasoline engine, large electric motor, water-wheel, wind-motor, water-motor, etc., etc., the "8-10" can be used to light a bunch of Tungstens, **to charge the biggest storage battery**, to electroplate, to run spark coils, etc. As a motor when run on 8 volts, the No. "8-10" will prove an exceedingly strong machine. We furnish this dynamo giving 8 volts and 10 amperes, that is 80 watts. With the No. "8-10" machine we were able to do the following: At 2,000 revolutions we lighted fifteen 8-volt 4 C. P. Tungsten lamps. At the same speed we also lighted six 6-volt 8 C. P. Tungsten lamps. The full 80 watts capacity is obtained at about 2,500 revolutions.

POINTS OF SUPERIORITY

Armature constructed of thin annealed electrical steel. Slotted for conductors. Coils are connected up in the usual drum manner. Laminations are keyed to shaft. Commutator pinned. Shaft unusually large for this size machine. 5/16 in. diam. at all points. Material—**steel ground** to size. Commutator same construction as on large dynamo. Hard drawn copper segments, heavily insulated with mica. Bearings "Non Grain" bronze, best obtainable. Extra large wick feeding oil cups. Retaining grooves prevent oil flying, and returns are provided so that excess oil returns to oil cup.

Field Magnets—Two pole, carefully bored to size. 12 part armature and commutator. Brushes—Two, made of special carbon metal, of square sections to prevent turning. Double brushes and double pole construction allows large brush contact surface with small commutator for the comparatively heavy currents encountered on low voltage work. Case—Cast iron, cast from die moulded patterns therefore no damaged castings are ever used. Winding—Shunt.

In designing this machine the very first consideration was for results. No endeavor was made to save copper, iron or workmanship. Instead of fancy finish the money was spent in result producing labor and materials. No weak end brackets, no skimpy brushes and brush holders, but good honest to goodness solid stuff that shows the value. It's a real bargain and one you will never regret investing in. The finish is black enamel with gold stripe. You will wonder how we can sell it at the price. Only the quantity makes it possible.

Machine is semi-enclosed, practically dustproof, a radical departure in small dynamo building. Pulley. 1 inch diameter, V-grooved for round belt.

Size over all is 7½x5½x5¼ inches.

Shipping weight 20 lbs.

No. AEX810 "Electro" 8 Volt 10 Amperes Dynamo, as described.   **$15.00**

Gentlemen:—   Reno, Nevada.
   I have received the No. 810 Dynamo order No. 258799. It is everything I expected and a little more.   RUSSELL L. BOARDMAN.

The "Electro" Hercules 12 Volt 9 Ampere Dynamo

100 WATT MACHINE

The latest acquisition to the "Electro" dynamo family is the Hercules 12 volt, 9 ampere generator here shown. It is a marvel of electrical and mechanical efficiency and simplicity. Good substantial design throughout characterizes this machine. But let us get down to brass tacks:

The ELECTRICAL FEATURES are: Shunt winding (best for charging storage batteries); laminated armature core to reduce eddy current losses to a minimum; field and armature terminals brought out to three heavy brass binding posts, mounted on fibre insulating block; especially heavy 16 segment copper bar commutator—mica insulated; special carbon copper alloy brushes of extra high conductivity—never run hot; brush holders of simple and thoroughly rugged design. Output at 2,000 revolutions per minute—12 volts, 9 amperes. Machine acts as motor when supplied with 12 volt, 9 ampere current developing nearly $\frac{1}{8}$ H. P.

The MECHANICAL FEATURES are: Good design of end-frame castings, giving ample ventilation to the electrical windings without exposing the working parts unduly to dust or damage; steel armature

No. BEX1209

shaft .5 inch in diameter; bearings of large size, fitted with large size wick oilers, the cups of which can be refilled without removal from the machine. The bearing houses are also designed with annular catch basins at each end to prevent oil from being thrown outward and return it to the cups. The shaft journals rotate in high grade phosphor bronze bearing sleeves, which may be removed for renewal by loosening a small screw; 1½ in. face, 1⅝ in. diameter crowned accurately turned cast iron removable pulley.

Size of dynamo—7 in. high x 11⅝ in. long x 6½ in. wide. Weight 30 lbs.
No. BEX1209 "ELECTRO" HERCULES DYNAMO. Price...... **$25.00**
Shipping weight 40 lbs.

The "Electroport" Dynamo

TYPE "SS"

This dynamo for a score of years has been one of the best articles that we have put before the discriminating experimenter. There is absolutely no machine that can come anywhere near our "Type SS" either in workmanship or output, considering its price. Our annual sales now amount to 3,000 machines—proof of its immense popularity.

The machine can be put to a variety of uses. When belted to a gas or other engine, or anywhere where there is surplus power, the type "SS" will light from 10 to 12 6-volt Tungsten lamps at a time. It will electroplate nicely up to a gallon plating bath and it will successfully charge small storage batteries. As a dynamo it may be run from 4 to 6 hours at a stretch and it will not be found to heat up unduly. When operated as a power motor it will develop an astonishing amount of power at six volts. Not more than 8 volts should be used. If desired to run as a motor through a transformer on 110 volts

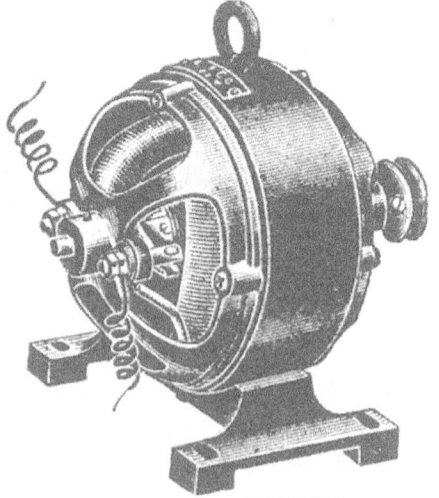

No. FX1331

A. C. field connections should be changed to series as per diagram furnished.

The "Electroport" Dynamo
(Continued)

The dynamo is semi-enclosed, after the latest design of large machines. The watt output (24) is sufficient for practical use; the price is within the reach of every experimenter. The armature is made from steel laminations, 1¾ in. diameter, six slots; with ¼ in. shaft. Commutator is of hard drawn copper, **ONLY MICA INSULATION USED.** The windings are dipped in insulating varnish and baked to guarantee the best results. The pulley is 1 in. diameter and grooved; for a 3/16 in. round belt; the machine may be connected to a gas engine, a sewing machine or other driving power. The dynamo is efficient, strongly made and handsomely finished in hard black enamel. It is equipped with an adjustable rocker-arm (not shown in cut) to adjust the brushes ensuring proper commutation and the best output of the dynamo. Size over all 5 in. x 5 in. x 5 in.

Approximate output at 3,000 revolutions: Open circuit 9 volts. Safe maximum load 6 volts, 4 amperes.

No. FX1331 Type "SS" Dynamo Complete. Shipping weight 7 lbs. **$6.00**

The "Electro" Hustler Motor

The motors we list below are the best American make on the market to-day. Built in a factory which has devoted itself to motor building since 1890. We guarantee each and every motor and will replace any proving defective of its own accord.

ACK100

This well-known motor, ever leading in efficiency and value, is a complete example of electrical science and workmanship. A very useful starting switch and the binding posts are mounted on the field to avoid disturbing the connections when it is desired to use the motor without the base. It is 3½ in. high, finished in black enamel with nickel-plated trimmings. Has a three-pole armature causing the motor to start without assistance when the current is applied. The ⅛ in. shaft is fitted with a pulley for running mechanical toys, models, etc., and drives a fan at a high rate of speed. Size over all 2½x3½x4 inches.

One dry cell will drive this motor at prodigious speed.

No. ACK100 "Electro" Hustler Motor, as described **$1.35**
Shipping weight 1 lb.

The "Electro" "O. K." Motor

It has taken experienced engineers many years of careful study to develop the highly efficient motors of the present day.

The O. K. is a very close copy in miniature, and its operation will be found pleasing and educational. It is an extremely powerful model. Every experimenter wants one at sight.

The armature is laminated, three pole, 1¼ inch diameter. Shaft over ⅛ inch diameter. Pulley 9/16 inch, grooved.

The finish is black enamel, with nickel plated trimmings. Size 4¼x4x4¼.

One dry cell will drive the motor or two cells when more power is required.

No. BCK179 "Electro" O. K. Motor, 2 to 4 volts, as described. Price **$2.30**
Shipping weight 3 lbs.

No. BCK179

Discharger

There is only one way to discharge a Leyden Jar and that is by using a discharger. Ordinary wire cannot be used, as it is impossible to draw a spark with pointed wires or other objects. The points act like lightning arresters and draw out the charge silently. By using our discharger the bright blue, crashing spark will jump between two balls as soon as one of the balls of the discharger touches the outside coating of the jar, while the other is brought close to the brass ball of the Leyden Jar. Nickel finished, hard rubber handle. Size over all 5x4 in.

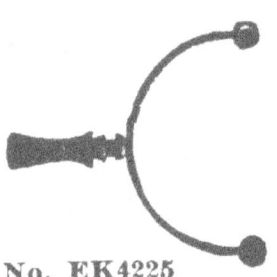

No. EK4225

No. K4225 Discharger, as described **$0.50**
Shipping weight 4 oz.

Induction Coil Core Wire

Number of pounds per complete core in decimal fractions of lbs.

This wire is used by us in all our spark coils and transformer coils. Annealed twice. The size is No. 22 B. & S. The wire is perfectly straight, machine-made and comes in 8, 10, 12 and 24 inch lengths.

Only sold by the pound.

No. BE529 Induction Coil Core Wire, per lb. **$0.25**

Length in Inches Diameter	8	10	12	24
½ inch	0.33...	0.41...	0.50...	....
⅝ "	.52...	.65...	.78...	....
¾ "	.75...	.94...	1.13...	2.25
⅞ "	1.02...	1.28...	1.52...	3.07
1 "	1.33...	1.67...	2.00...	4.00
1⅛ "	1.69...	2.11...	2.53...	5.06
1¼ "	2.08...	2.60...	3.13...	6.25
1⅜ "	2.52...	3.15...	3.78...	7.56
1½ "	3.00...	3.75...	4.50...	9.00
1¾ "	4.08...	5.11...	6.13...	12.30
2 "	5.33...	6.67...	8.00...	16.00
2¼ "	6.75...	8.44...	10.10...	20.30
2½ "	8.33...	10.40...	12.50...	25.00
2¾ "	....	12.60...	15.10...	30.30
3 "	....	15.00...	18.00...	36.00

DIMENSIONS AND WEIGHTS OF IRON WIRE CORES:

EXAMPLE.—A core is desired 12 inches long, 1½ in. diameter. Referring to table above we find at the intersection the weight 4.50. (This is 4.50 LB., not $4.50.) Now multiply 4.50 lbs. with 25c. (price per lb.) and we have

4.50 × .25 = $1.12½—price of core.
Shipping weight of any of the above per lb., 2 lbs.

Friction Insulating Tape

NO. BK1587

We carry a high grade black Tape known throughout the country. Quality of this tape is guaranteed. **WE DO NOT CARRY THE CHEAP WORTHLESS KIND** that does not stick.

No. BK1587 ½ lb. Roll. Size 4½ in. diam, ¾ in wide **$0.20**
Shipping weight 1 lb

Gentlemen:—   Brighton, Mich.

I received my Static Machine. IT WORKS FINE, GIVING A FULL 3-INCH SPARK You will find 4c. in stamps enclosed for one of your catalogues.   Yours very truly,   E. R. ROBERTS.

Making Selenium Cells

BY HOMER VANDERBILT

Extract from the September, 1916, issue of
The Electrical Experimenter
Published by Permission

One of the simplest forms of selenium cells is the Bidwell type, which consists of a flat, insulated sheet, wrapped with two separate bare wires in a single layer, each of which is insulated from the other, Fig. 1. The insulating sheet consists of a small piece of mica. The size of the sheet depends upon the size of the selenium cell, but a piece of mica measuring 2½x1 in. is a convenient size. It will be found that such a cell is suitable for practically all sorts of work, such as in the transmission of photographs over a wire, and in television, where large flat cells prove very effective. Two No. 30 B. & S. bare copper wires are wound closely about this mica as shown in the sketch. Extreme care should be taken in keeping the two wires separated from each other and at the same time keeping them very close. If the two wires were widely separated, the resistance of the selenium cell would be very high and thus the sensitivity of the cell would be decreased. A good method of keeping the wires closely spaced and at the same time not short-circuiting them, is to make a number of grooves with a knife in the edges of the mica sheet, which must be equally spaced and in which the wires are wound. The diagram clearly shows how it should be made.

When the skeleton of the cell is made the next and very important operation is the application of the selenium to the wires and to render this material sensitive to light. This last operation is called annealing. The process of annealing is vitally important, as the sensitiveness of the finished cell will depend upon the process.

In order to perform this operation successfully, the following apparatus is required: a stand A, Fig. 2, having a 6-inch ring B, and a holder E, in which a 200 deg. C. thermometer D, is placed. A standard form of laboratory Bunsen burner must also be obtained. The apparatus should finally be arranged as observed. The next step is to apply the selenium, which must be chemically pure. The selenium must be applied to the skeleton of the cell as follows: Place the form on the ring stand as illustrated in Fig. 2 and heat it with the Bunsen burner until the selenium will melt when brought to the surface. It should not be heated higher than 212 deg. C. Several drops of selenium should be put on the wire grid, and with the aid of a knife blade distributed equally over the complete grid area. Care should be taken to make the selenium surface very thin; in fact, it should be almost, and if possible quite, transparent. Having done this, the unfinished cell is allowed to cool slowly.

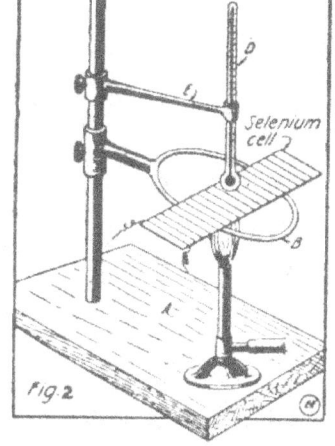

Fig. 2

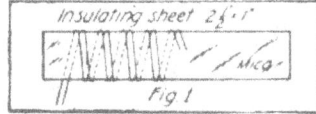

Insulating sheet 2½·1   mica   Fig. 1

We now come to the annealing of the selenium. This may be accomplished by placing it on a strip of mica under which the Bunsen burner is placed. The flame is slowly increased until the surface of the selenium turns a dull gray color. The flame should not be increased after the first signs of melting appear. If melting is observed, the burner must be quickly removed and the flame reduced. The dark gray spots will harden in a few seconds, after which the flame should be reduced and left for two to three hours with the temperature just below, but never above, the melting point of the selenium. The annealing process is then completed by allowing the cell to cool very slowly as the flame is gradually lowered and finally extinguished.

The resistance of each cell depends upon the manner in which it is built, so that no definite statement can be made as to what the resistance of the cells will be.

[Contrary to general opinion selenium in its pure state has an enormously high resistance and can not be used on a cell grid by simply melting it on. It must be crystallized by annealing to render it conductive. The resistance of a stick of C.P. selenium is practically infinite.

Selenium

This peculiar substance is a conductor of electricity while exposed to light rays. An insulator in the dark. Used to make the well-known Selenium Cells. Will close a relay when match is lighted near cell.

Selenium will solve many problems during this century. It is one of the most wonderful substances ever discovered. The selenium we handle is the very highest grade obtainable for the making of selenium cells. It is exactly the same quality as is used in our own cells and crystallizes very readily. This power to crystallize readily is very important as on it the speed and actual working of a selenium cell depends. Some very successful experiments have been conducted with selenium as a potentiometer and its high resistance renders it particularly well suited for such instruments. Remember we guarantee our selenium to be chemically pure not commercially pure.

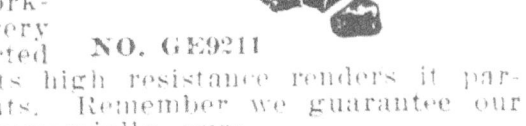

NO. GE9211

No. GE9211   Selenium Metal (Chemically Pure), per oz..........   **$0.75**
     Shipping weight 2 oz.
No. CK9211a   One-quarter ounce (smallest quantity sold)......   **$0.30**
     Shipping weight 2 oz.

Bunsen Burners

No. FK1352

This useful appliance is mounted on iron base, with Stop to regulate the mixture of air. The Brass tip inside of tube can be adjusted to regulate the flow of gas. When used with ordinary illuminating gas gives intense heat. Size 5½ x2 in.

No. FK1352   Bunsen Burner, as per illustration, each **$0.60**
   Shipping weight 1 lb.

Glass Spirit Lamp

Invaluable to the experimenter. Used in a thousand different ways, to bend glass rods and tubing, to solder wires, etc. Uses wood alcohol. Size 3½ x2 in.

No. BE1339   Spirit Lamp, as per illustration, each ....

NO. BE1339

$0.25
Shipping weight 1 lb.

Dear Sirs:—
The Electro Whistle you sent me is all you claim for it. I connected it up and use it on my bicycle for a "Claxon." I only use two batteries.
     KIRK S. PATRICK.

A complete chapter on "HOW TO MAKE THINGS" is contained in the "EXPERIMENTAL ELECTRICITY COURSE" in 20 lessons which is given FREE with one year's subscription to the "Electrical Experimenter Magazine." See announcement on back cover of the catalog.

SKINDERVIKEN Transmitter Button

THE SKINDERVIKEN TRANSMITTER BUTTON presents the latest advance in microphones, and marks a revolution in transmitter construction. It works on an entirely new principle, takes up practically no room, and marks the end of all telephone transmitter troubles.

The SKINDERVIKEN TRANSMITTER BUTTON can be placed in any position and it will talk loud and distinct, being at the same time extraordinarily sensitive. It was primarily designed to replace the old damaged or burnt-out transmitter. Simply unscrew and remove the telephone transmitter front, disconnect the two inside wires, unscrew and remove the bridge and the old electrode. There remains only the diafram. These wires are then connected with the Skinderviken button, the latter screwed to the diafram, and after screwing the old transmitter housing together again, the telephone is ready for work.

EXPERIMENTERS will be particularly interested in different experiments that can be performed with the Skinderviken Button. Fig. 1 shows the Skinderviken button attached to the back of an Ingersoll watch case. When speaking towards the inside of the case, it will be found that the voice is reproduced clearly and loudly. Fig. 2 shows another interesting stunt. By attaching the button to a tin diafram about the size of half a dollar, and by holding the diafram at the side of the throat, as shown, speech can be transmitted with surprising clarity. Fig. 3 shows an interesting stunt, whereby a hole is drilled in the side of a thin glass

Fig. 1   Fig. 2   Fig. 3   Fig. 4   Fig. 5

water-tumbler; the sides of the glass thus acting as a diafram, the voice is clearly transmitted. Fig. 4 shows how to transmit phonograph music at a distance merely by drilling a small hole in the phonograph arm and attaching the Skinderviken button, a very favorite experiment with all experimenters. Fig. 5 shows how a very sensitive Detectophone can be made by placing one of the buttons in the center of a lithographed cardboard picture, so that only the small brass nut shows. The large surface of the picture acts as a big diafram and the voice is well reproduced.

We have such unlimited confidence in the Skinderviken transmitter button that we make the following remarkable offer: Send us one dollar ($1.00) for which we will mail one button prepaid. If for any reason whatsoever you do not wish to keep the button, return it within five days and your money will be refunded.

No. 440—Skinderviken Button, prepaid **$1.00**

Electro Gold Leaf

This substance, which is not pure gold but composition gold leaf, nevertheless is very much better than standard gold leaf, for the reason that it can be handled with the fingers with ease without adhering thereto. Ordinary gold leaf, as is well known, sticks to the fingers and tears remarkably easy. The foil which we are marketing is even thinner than ordinary gold leaf. It is used mainly for electroscopes and for other experiments where an extremely thin metal foil is required. This metal leaf is so thin that if held against the light, the light will shine through it the same as through standard gold leaf. Many experiments can be performed with this foil, and it is also used for very thin radio condensers in connection with extra thin "onion" paper to build up condensers of enormous capacity having but little bulk. This material is only sold in sheets measuring 4¾x5 inches, not less than a dozen sheets being sold. Sheets come packed with paper between each metal foil, and we guarantee satisfaction.

No. 666 Electro Gold Leaf as described, twelve sheets, prepaid.. **$0.50**

The "Electro" "Detectiphone"

FOR AIDING THE DEAF; RADIO AMPLIFIERS AND DETECTIVE WORK

We present herewith the latest improved pattern of a super-sensitive telephone set, belonging to that class of electrical devices commonly known as **Detectiphones**. This instrument is marvelously sensitive to any and all audible sound waves and will pick up and articulate or reproduce in its receiver the faintest speech or whisper several feet away from the transmitter.

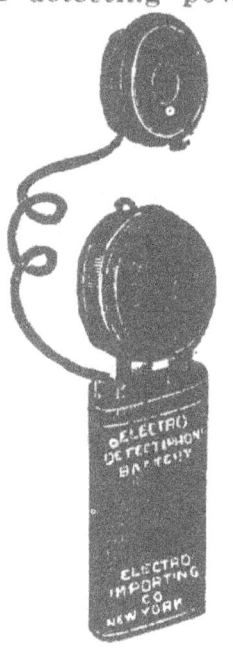

NO. AEX9750

Briefly it comprises a specially constructed and accurately made microphonic transmitter, together with a specially adapted watch-case receiver, connected with suitable connecting cord, to a battery. The whole forms in reality a miniature telephone set of super-sensitive sound detecting power.

The instrument is, firstly, of enormous benefit to all those who are afflicted with partial deafness, due to catarrhal stoppages, etc. It may be worn inconspicuously by any lady, as the transmitter can be placed under a shirtwaist, etc., and the connecting wire led up inside the collar, to the small watchcase receiver at the ear, held in place by a narrow black headband, and the receiver itself is very easily covered over by the hair, making it invisible.

A small switch is placed on the receiver to cut in the instrument, whenever conversation is to be carried on. It is usually switched out as soon as the conversation is ended. It is well to leave the instrument switched in where there is busy traffic, etc., so that all noises can be readily perceived, which is a great boon to those hard of hearing. The batteries, of which there are two furnished with each "Detectiphone," are specially made and will last several months with common usage. Of course the more the instrument is used, the more drain on the batteries.

Radio signals can be amplified by a suitable arrangement of these super-sensitive "Detectiphones" and one of the most successful commercial telephone and radio amplifiers, employs this system of stepping up the strength of the signals or sounds. In general, a

NO. AEX9750

high resistance radio telephone receiver is placed close against the transmitter of the "Detectiphone," and both are bound around the edge with tape, to make the space between them sound-proof. The incoming wireless signals will thus be transferred from the radio receiver to the transmitter and thence to the special watch-case receiver of the "Detectiphone." By employing two or more of these instruments, the signals can thus be stepped up in successive stages. A common arrangement used for the purpose makes use of three stages, necessitating, of course, three instruments. A large field is opened here for experiment to the amateur radio enthusiast.

Detectives have been using these wonderfully sensitive telephone sets for several years now, and they are very necessary to any Detective or Agency engaged on difficult cases. The transmitter is placed behind a

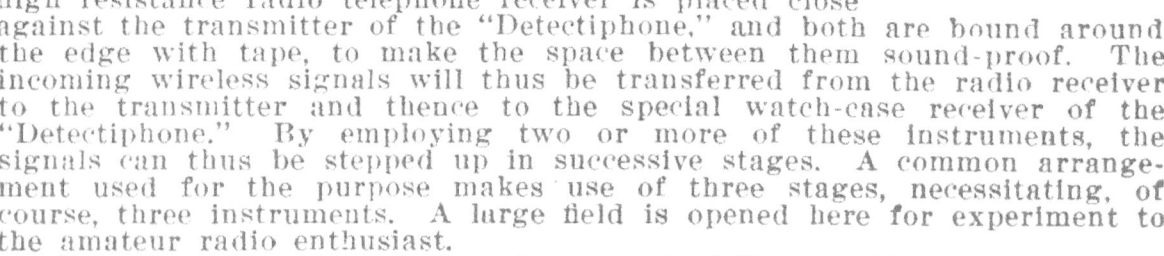

A complete chapter on "X-RAYS" is contained in the "EXPERIMENTAL ELECTRICITY COURSE" in 20 lessons, which is given FREE with one year's subscription to the "**Electrical Experimenter Magazine.**" See announcement on back cover of catalog.

The Electro Detectiphone (Continued)

picture on the wall, or in some other concealed location, and by extending the length of the cord between the transmitter and receiver, the detective or operator can be located as far as 200 feet from the transmitter. Flexible twin-conductor is used for thus extending the circuit.

These instruments are very accurately made and extremely well finished in black enamel. Furnished complete, in neat portable leatherette case 2x2½x7 in., with transmitter, receiver, head-band and two batteries. Supplied complete only.

No. AEX9750 "Detectiphone." Price **$15.00**

Shipping weight 2 lbs.

No. DE9751 Extra Batteries. (Shipping weight 1 lb.) **$0.45**

No parts sold separately

"Electro" Loud-Talker

We present herewith two little instruments for which we have had a long and persistent demand.

While the **Detectiphone** which we list in our catalog is a very high-grade instrument and adapted for all kinds of special work, we realize that its price is perhaps above what the average Experimenter and Dabbler wishes to spend, and for this reason we have originated this little set for strictly experimental uses.

This outfit has been gotten up solely for the Experimenter and for this reason we are selling it "Knocked Down." In other words, the instruments **come all ready for you to assemble,** all the parts, screws, nuts, washers, etc., being furnished. Complete directions how to assemble accompany each set. With a pair of pliers and a screw-driver, the outfit can be readily put together in less than twenty minutes.

The most important point is that the telephone receiver spool comes already wound complete, and the Experimenter will, therefore, not need to wind his own spool.

The outfit when assembled comprises a highly sensitive **CARBON BALL MICROPHONE** with carbon diaphragm of exactly the same type as is used with our **Detectiphone.**

The "Back Plate" which holds the carbon balls has five cup shaped polished depressions, each accommodating about twelve to fifteen of the special carbon balls furnished in a bottle.

CEK205

The receiver is our No. 1024 style with the difference that no magnet is used in the same for the reason that the function of this instrument is electro magnetic, the same as all loud-talking phones.

The spool is wound with special enameled wire for five ohms, standard with our **Detectiphone.**

This instrument works best on two dry cells, and particular attention is called to the fact that in order to work, the loud-talker requires a fairly heavy current and for that reason thick wires must be used for connecting the transmitter with the loud-talker. If this is not done, the voice will be weakened considerably. If no heavy wire is at hand, more batteries must be used to compensate.

With this instrument **no switch is required;** if one is through talking all that is necessary is to lay the transmitter **face up,** which automatically cuts out the current.

USES: This instrument can be used to transmit phonograph music from one room to another; used as a Detectiphone; as a Radio Amplifier; as a telephone extension (by placing the regular telephone receiver against the sensitive transmitter); as a "Howler" (Whistling Micro-telephone); dictating to stenographer at a distance; for salesmen to talk "through" window (Loud-Talker outside in street, microphone transmitter for salesman, talking into same); for restaurants for talking to the chef, and a hundred other uses. Many young experimenters are developing a lucrative business selling this appliance to various merchants at a good profit.

Outside of the two instrument parts, one three foot cord is furnished with the sensitive microphone as shown. Blueprint, instructions, etc., are furnished.

No. AEK204 "Electro" Loud-Talker Outfit Parts "Knocked Down," complete........ **$1.50**

No. CEK205 "Electro" Loud-Talker Outfit, same as above except that it is already **$3.50** assembled and tested at factory. Set complete ....

Shipping weight 1 lb.

The "Electro" High Frequency Apparatus

VIOLET RAY MACHINE

We can now present for the first time a piece of apparatus which has been wanted by physicians, as well as laymen, for years, and is invaluable in every home. It is a portable, high frequency outfit with all the therapeutic value of the large machines you see in some physicians' offices. This apparatus will work on either direct or alternating current of 110 volts, but not on batteries.

The "Electro" Violet Ray outfit delivers the true high frequency current that is of so much value. It is almost noiseless in operation. **ABSOLUTELY SAFE**, as no painful spark can be drawn by the person using the machine. The weight, being only 1¼ lbs., makes it positively ideal for portability. In use the apparatus consumes less than ¼ of the current consumed by a 16 c.p. incandescent lamp. It is truly remarkable

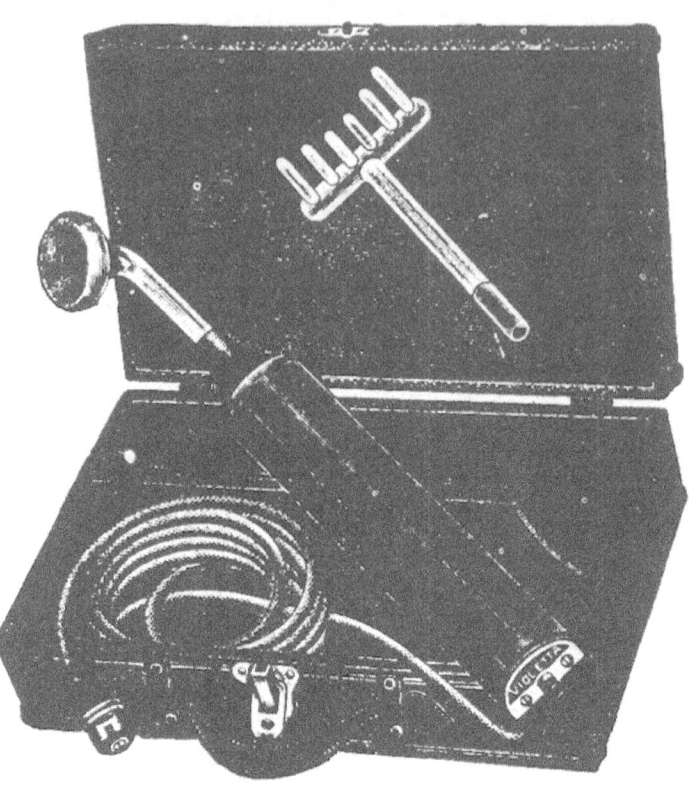

No. AHEK1571

to note the large amount of ozone given off by this machine.

The vibrator spring has been the subject of much investigation and the one used gives a perfect series of high speed interruptions. The contact points are the result also of continuous experiment and can be used for years without burning out. The entire apparatus is enclosed in a molded hard rubber case, eliminating all possibility of shock and the cord and attachment plug furnished are as substantial as can be obtained. With each outfit a No. AEK1577 General Electrode, as shown, is furnished. Every piece of apparatus is carefully tested before leaving the factory and will do perfect work.

High frequency apparatus has one great advantage over all others. While operating it gives off a gas technically called ozone. This gas is the greatest disinfectant that is probably known to-day, killing microbes on contact. Ozone also acts as a stimulant which you probably have experienced yourself after a thunderstorm. The sharp, indefinable smell so apparent after a thunderstorm, which makes you feel so refreshed and light, is simply ozone produced by the lightning discharges between the earth and clouds. This capacity of high frequency apparatus is used very frequently for purifying the air of overcrowded rooms. Our "Violet Ray" machine will purify the air in a small room in a remarkably short time by the large volume of ozone it generates.

Thus our Electro "Violet Ray" machine will not only be of service in the definite manner in which you desire to use it, but will, in addition, produce a quantity of ozone which will refresh and invigorate you, kill microbes and clear the air.

Complete directions and instructions are furnished. Every home that has electric current should have an Electro Violet Ray Apparatus. It is positively invaluable for treatment of skin diseases, nervous disorders, etc., etc. Size 11x2 in.

The "Electro" High Frequency Apparatus.
(Continued)

No. AHEK1571 "Electro" High Frequency Apparatus, as described, with one No. AEK1577 Electrode, and case, complete.... **$18.50**

No. AFE1572 "Electro" High Frequency Apparatus-Carrying Leatherette Case only for Electrodes............... **$1.65**

Shipping weight 3 lbs.

No. AEK1577 Renewal General Electrode...................... **$1.50**

Shipping weight 2 lbs.

* The machine can also be used for purification of water by the ozone method and many other experiments in H. F. work familiar to the experimenter.

It is truly remarkable how many ailments this apparatus will successfully, either heal permanently or bring at least a wonderful relief. Boils and pimples, for instance, will come to a head astonishingly quick, while a sick headache or a backache, will positively be relieved within 10 minutes. The current of the apparatus is absolutely harmless.

Electrodes for Violet Ray Outfit

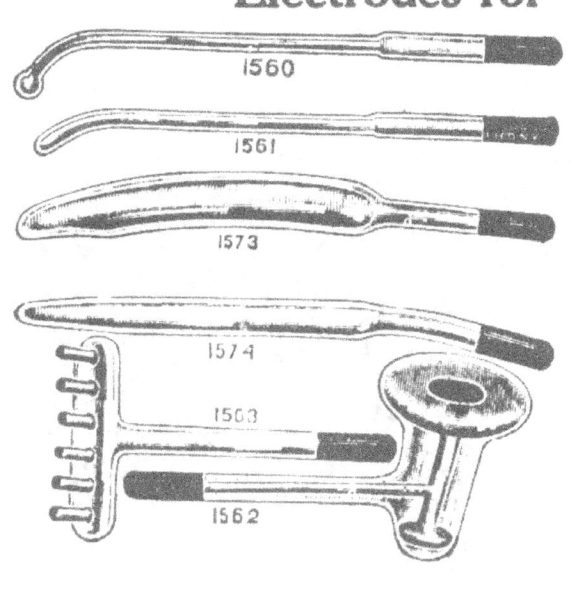

These electrodes are for special purposes as listed below. Every electrode is made of special imported glass carefully annealed and will stand the strains due to contraction and expansion from heat causes without the slightest danger of cracking. All electrodes are protected by a metal cap which doubles the life of the electrode. No. ABE-1560 and No. ABE1561 are insulated tubes which are of advantage in so far as current may be introduced without loss into the orifice of the body. In using a plain electrode for an orifice much of the current is lost at the point of contact with the body. Electrodes should always be inserted before the current is turned on and then turned off before the tube is removed.

Plain Electrodes

No. CX1562 Condenser Electrode condenses the current and produces a strong, even flow of current, generating extreme heat which is very desirable in deep seated cases........ **$3.00**

No. BX1563 Comb Rake Electrode, used for scalp treatment, falling hair, dandruff, gray hair and for stimulating the scalp cells **$2.00**

No. GE1573 VAGINAL ELECTRODE **$0.75**

No. GE1574 RECTAL ELECTRODE **$0.75**

Insulated Electrode

No. ABE1560 INTERNAL THROAT ELECTRODE. Very extensively used for treatment of tonsilitis, hypertrophy of the tonsils, ulcers of the tonsils, etc., etc.................... **$1.25**

No. ABE1561 NASAL AND EAR ELECTRODE. A form unusually successful for treatment of rhinitis, nasal catarrh, etc. **$1.25**

Shipping weight, all sizes, 2 lbs.

Above Electrodes to be used only with No. AFGE1571 Violet Ray machine.

"Electro-Therapeutics"

A complete chapter on "ELECTRO-THERAPEUTICS" is contained in the "EXPERIMENTAL ELECTRICITY COURSE" in 20 lessons, which is given FREE with one year's subscription to the "**Electrical Experimenter Magazine.**" See announcement on back cover of catalog.

CONNECTORS

No. AE6323

These connectors insure reliable connections and avoid the annoyance experienced when using ordinary wire. Made of specially prepared copper, just the right length to connect various sizes of dry cells.

No. AE6323 Connectors, per dozen **$0.15**

Shipping weight per doz. 4 oz.

Electro-Magnets

We have had a persistent demand for small electro-magnets such as are used in bells, buzzers, etc., and we decided to list them separately.

The electro-magnets which we show here are carefully built and carefully wound, all material going into them being high grade. Coil heads are smooth fibre, the cores being a soft annealed iron. Resistance 3 ohms.

The two electro-magnets as per illustration are mounted on an iron plate; these two little magnets are quite powerful, lifting 5 to 6 pounds on three cells. They are wound neatly with green wire and are particularly recommended for making small electric engines, all kinds of lifting experiments, to make Radio buzzers, bells, annunciator drops, as well as making all kinds of other experiments in which a good, powerful, but small electro-magnet is required. The sizes over all are 1⅝ in. x 1¼ in. x ⅞ in. Each spool measures over all 1⅛ in. x 1¾ in. Coil heads measure 1¾ in. over all. Size of core is ⅝ in. in diameter.

DK401

No. DK401 Set of two electro-magnets, mounted as described, pair **$0.40**
Shipping weight 1 lb.

No. BK402 Single electro-magnet, as described, each............... **$0.20**
Shipping weight 4 oz.

Gentlemen:—   Bakersfield, Cal.

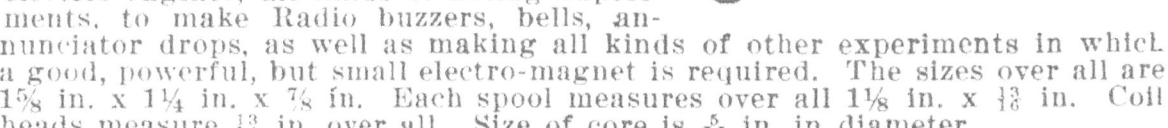

Your articles are A1 and YOUR PRICES LEAVE NOTHING TO BE DESIRED by the amateur with a short pocket-book. I have set up a couple of the small meters purchased from you some time ago, on a small switchboard, and besides giving the BEST OF SERVICE, they add very materially to the APPEARANCE of the other apparatus. The ½-inch coil and tubes purchased of you OVER A YEAR AND A HALF AGO STILL GIVE FINE DEMONSTRATIONS.

Yours respectfully.   O. BICKERDIKE.

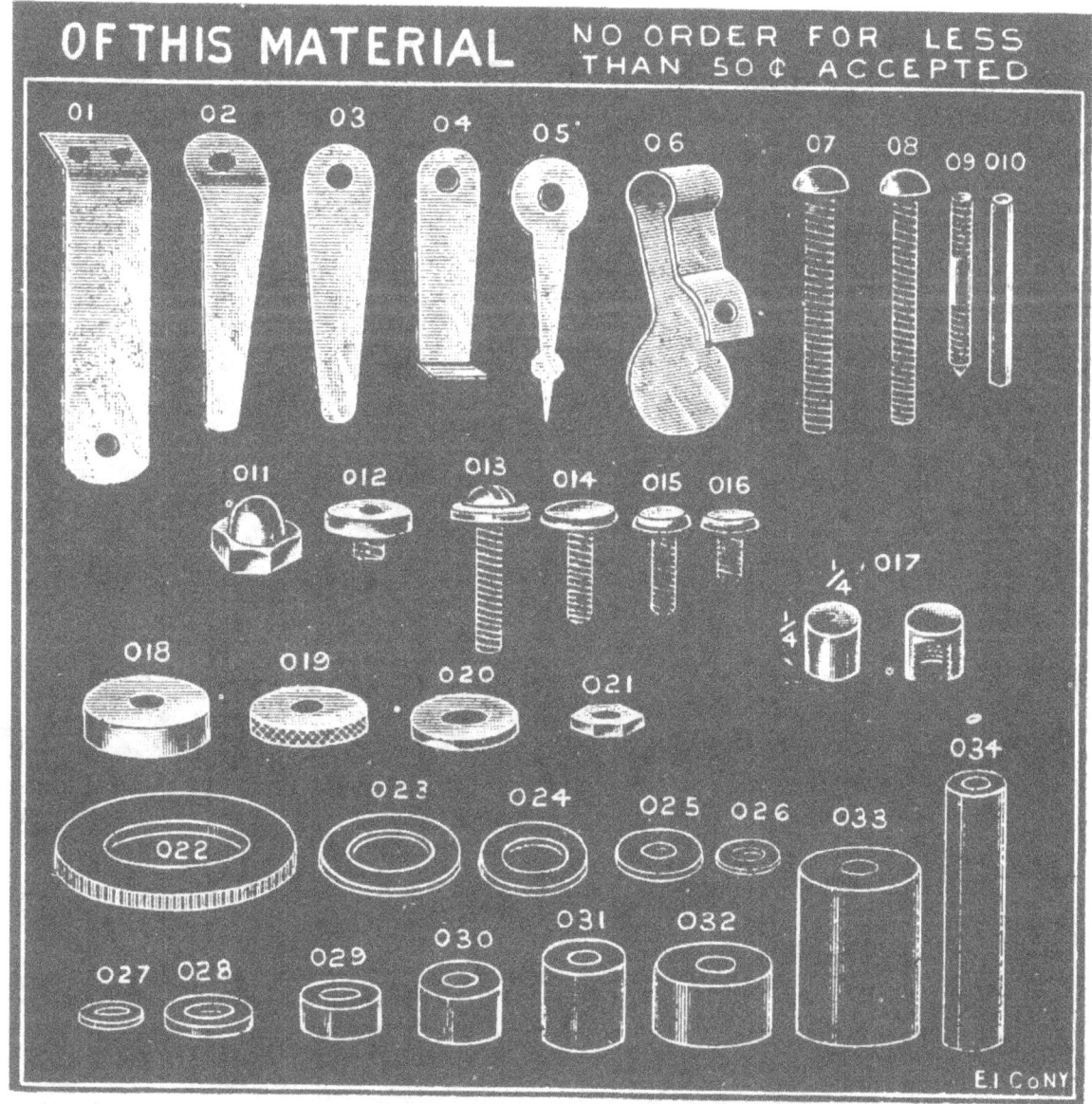

OF THIS MATERIAL   NO ORDER FOR LESS THAN 50 ¢ ACCEPTED

No. 01   Brass Leaf Detector Spring. 2⅜ in. long, ½ in. wide, 1/64 in. thick. Three holes pass 8-32 screw. **Price each, $0.10.**

No. 02.   Stiff Brass Spring. 2⅜ in. long, ⅝ in. wide, 1/32 in. thick, 5/16 in. wide at top. Hole passes 8-32 screw. **Price each, $0.10.**

No. 03.   Nickel Switch Blade. 2 in. long, 1/32 in. thick. Hole passes 8-32 screw. **Price each, $0.10.**

No. 04.   Brass Nickeled Radio Switch Blade. 1½ in. long, turned over end ⅛ in. Hole passes 8-32 screw. End ⅜ in. wide. **Price each, $0.06.**

No. 05.   Brass Pointer, Nickeled, Polished. 1¾ in. long. Hole passes 8-32 screw. **Each $0.06.**

No. 06.   Brass Detector Cup Spring. Top of round part takes cup. 1½ in. long. Spring stock. Hole passes 8-32 screw. **Each $0.10.**

No. 07.   Slotless Screw. Head nickeled. 1¾ in. long. 8-32. **Price each, $0.03.**

No. 08.   Slotless Screw. Head nickeled. 1½ in. long. 8-32. **Price each, $0.03.**

No. 09.   Brass Pointed Screw. 8-32 thread as shown. 1⅛ in. over all. **Price each, $0.02.**

No. 010.   Plain Steel Rod. ⅛ in. diameter, 1⅛ in. long. **Price each, $0.01.**

No. 011.   Brass Nickel-plated Hexagon Cap Nut. ⅜ in. high, ⅜ in. diameter, tapped 8-32 thread. **Price each, $0.02½.**

No. 012.   Steel Contact Piece. Silver contact in center. ⅜ in. diameter, ¼ in. high over all, ⅛ in. thick. Thread 8-32. **Price each, $0.08.**

No. 013.   Washer Screw, Iron. 1 in. long. 8-32 thread. Washer ⅜ in. diameter. **Doz., $0.10.**

No. 014.   Switch Point. Large, brass, nickeled. ½ in. under head, ⅜ in. diameter. **Price each, $0.02.**

No. 015.   Switch Point. Brass, nickeled, 7/16 in. under head, 8-32 thread, diameter 5/16 in. **Price each, $0.01.**

No. 016.   Brass Nickeled Switch Point. ¼ in. under head, 5/16 in. diameter, 8-32 thread. **Price each, $0.01.**

No. 017.   Radio Switch Point. Brass, nickel plated. ¼ in. diameter, ¼ in. high, tapped 6-32 thread. **Price doz., $0.15.**

No. 018.   Heavy Brass Washer. 9/16 in. diameter, 3/16 in. thick. Hole passes 8-32 screw. **Price each, $0.04.**

ALLOW A SUFFICIENT AMOUNT FOR POSTAGE

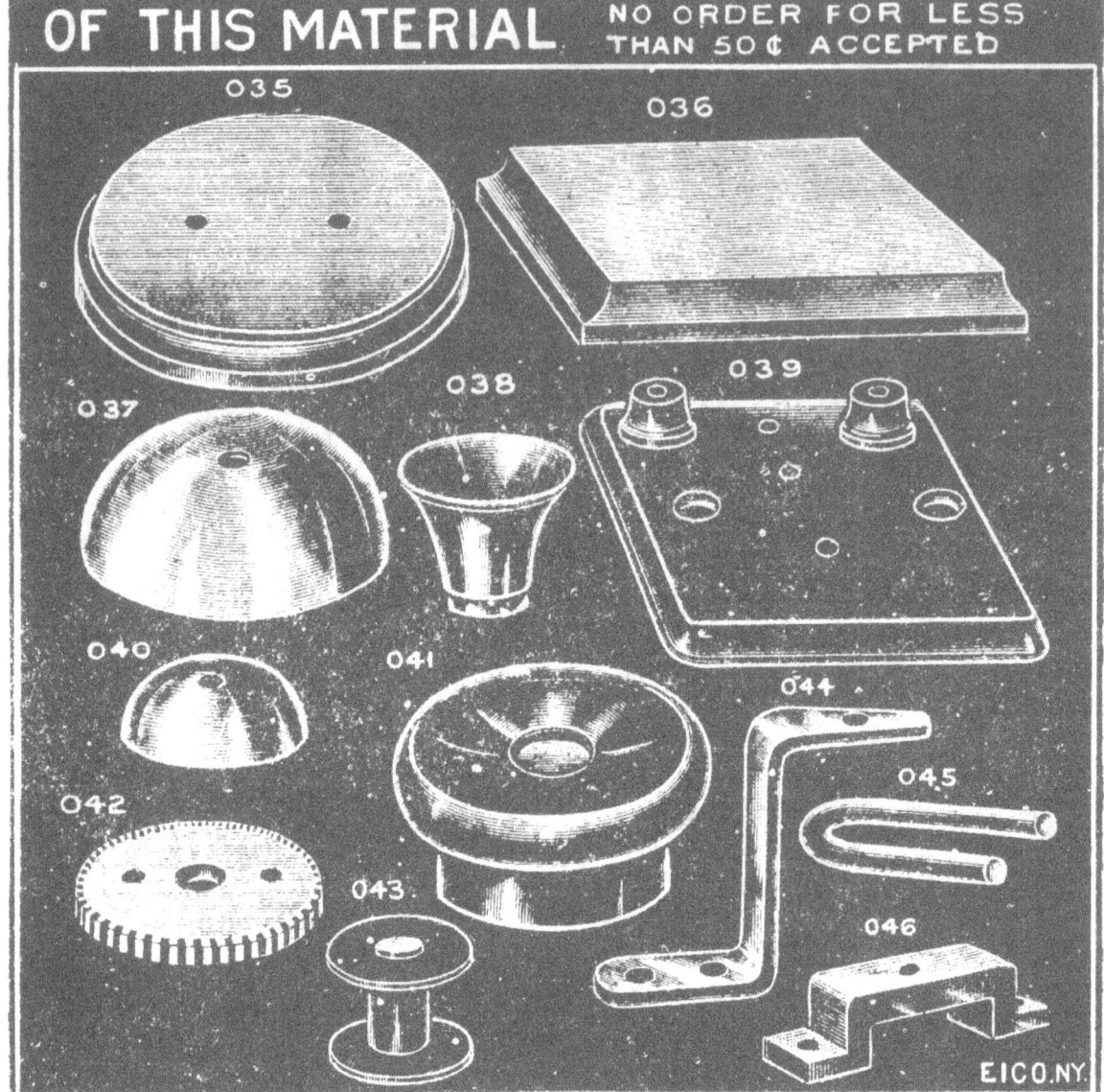

OF THIS MATERIAL   NO ORDER FOR LESS THAN 50¢ ACCEPTED

No. 019. Knurled Brass Piece. 9/16 in. diameter, ⅛ in. thick. Hole tapped 8-32 but does not pass all the way through. Used for Spark Gaps, Detectors, etc. **Price each, $0.03.**

No. 020. Brass Washer. Diameter ½ in., 3/32 in. thick, hole 3/16 in. **Each, $0.01½.**

No. 021. Brass Hexagon Nut. 8-32 thread, 3/64 in. thick, diameter ⅜ in. **Dozen, $0.12.**

No. 022. Knurled Fibre Ring. For Detector Cups, etc. 1¼ in. diameter, ⅛ in. thick, hole 13/16 in **Price each, $0.06.**

No. 023. Fibre Washer. 15/16 in. diameter, 1/16 in. thick, hole 5/16 in. **Each, $0.02.**

No. 024. Fibre Washer. 13/16 in. diameter, 3/32 in. thick, hole 5/16 in. **Each, $0.01.**

No. 025. Fibre Washer. ⅝ in. diameter, 1/16 in. thick, hole 3/16 in. **Doz., $0.10.**

No. 026. Fibre Washer. ⅜ in. diameter, 1/16 in. thick, hole 3/16 in. **Doz., $0.08.**

No. 027. Thin Fibre Washer. ⅜ in. diameter, 1/64 in. thick, hole 5/32 in. **Doz., $0.06.**

No. 028. Fibre Washer. 7/16 in. diameter, ⅛ in. thick, hole 5/32 in. **Price doz., $0.10.**

No. 029. Fibre Piece. ¾ in. diameter, ¼ in. high, hole ⅜ in. **Price each, $0.03.**

No. 030. Black Wood Piece. 5/16 in. high, ¾ in. diameter, hole passes 8-32 screw. **Each, $0.01½.**

No. 031. Fibre Piece. ½ in. high, ½ in. diameter, hole passes 8-32 screw. **Each, $0.03.**

No. 032. Fibre Piece. ½ in. high, ⅝ in. diameter, hole passes 8-32 screw. **Each, $0.05.**

No. 033. Wood Piece. 1-1/16 in. high, ⅝ in. diameter, hole passes 8-32 screw. **Each, $0.05.**

No. 034. Black Ebony Enameled Wood Pillar. 1¾ in. high, 7/16 in. diameter, hole passes 8-32 screw. **Price each, $0.05.**

No. 035. Wood Instrument Base. Mahogany imitation finish. 3½ in. diameter, 1½ in. thick. Two holes passing 8-32 screw 1½ in. apart. **Price each, $C.10.**

No. 036. Wood Instrument Base. 3½ in. x 2¼ in. ½ in. thick. Beveled edges, imitation mahogany finish. **Price each, $0.10.**

No. 037. Steel Nickel Bell Gong. 2½ in. diameter. **Price each, $0.05.**

ALLOW A SUFFICIENT AMOUNT FOR POSTAGE   (OVER)

(Continued from preceding page)

No. 038. Brass Nickel Plated and Polished Horn. 1½ in. diameter, 1¼ in. high, lower hole ⅝ in. To be put on telephone caps for loud-speaking telephones (ream out cap hole and turn horn rim over). **Price each, $0.15.**

No. 039. Hard Rubber Composition Detector Base. 3 in. x 3 in., ¼ in. thick. Made to fit No. 044 standard. **Price each, $0.20.**

No. 040. Steel Nickel Bell Gong. 1½ in. diameter. **Price each, $0.03.**

No. 041. Telephone Receiver Case and Cap with Diaphragm. As used in all our telephone receivers. For phones, transmitters, etc. **Price complete, $0.25.**

No. 042. Gray Horn Hard Fibre Driving Gears. Teeth accurately machined. 1⅝ in. diameter, ⅛ in. thick. Center hole ¼ in. **Price each, $0.15.**

No. 043. Telephone Spool. With two 15/16 in. fibre heads and iron core, tapped at bottom for 6-32 thread. Fits 041 telephone receiver shell. Wind your own receivers. **Each, $0.10.**

No. 044. Nickel Detector, Standard. 1⅞ in. high over all. Fits our 039 base. Top hole tapped for 8-32 screw. Two lower holes pass 8-32 screw. **Price each, $0.20.**

No. 045. Iron Horse-shoe Magnet. Nickeled. 1⅞ in. long, rod ¼ in. diameter. **Each, $0.10.**

No. 046. Brass or Aluminum Spark Coil Vibrator Bridge. See No. 047. 2 in. long, ½ in. wide, ½ in. high; center hole tapped, two side holes pass 8-32 screw. **Price each, $0.20.**

No. 047. Large Brass Knurled Vibrator Screw. To fit above, with spiral check spring. Large ⅛ in. diameter platinum contact for 1 in. spark coils. **Price each, $0.50.**

Magnetic Compass

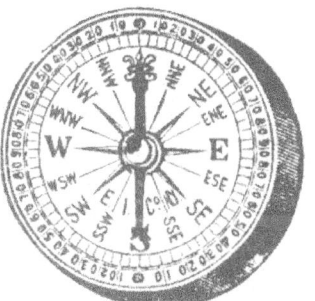

No. BE1325

This fine compass which should not be judged by its low price is an exceedingly accurate instrument.

Each instrument is warranted to be accurate. Its needle will always point NORTH, and it therefore is invaluable for orientation purposes. It is extremely sensitive. The magnetic action of an electric street car a block away will deviate the needle. The metal casing is of polished brass.

Instructions tell how to use this instrument for orientation purposes, telling polarity of magnets, tells if an electric wire is "alive" or "dead," making a voltmeter with this compass, etc., etc.

Size 1¾ in. diameter, ½ in. thick.

No. BE1325 Magnetic Compass, as described, with directions **$0.25**

Shipping weight 4 oz.

Connectors

NO. CE7590

These handy connectors are used a good deal to make any sort of temporary connection and their use will be plain to everybody. Used especially to connect Wireless Head Sets to the instruments. When through receiving messages, pull connector apart and put phones away. There are, of course, hundreds of other uses such as for connecting portable lights, chandeliers, small motors and any temporary connection which must be made and unmade quickly. The connector is very substantially made and will last a lifetime. It will carry from 3 to 5 amperes continuously without heating. You should have one around at all times just for emergency work.

No. CE7590 Separable Connector, complete, each................ **$0.35**

Shipping weight 4 oz.

The "Electro" Soldering Outfit

No. DK1144

Contains soldering iron, scraper and bar of finest solder, one box soldering salts, and full directions; all enclosed in handsomely finished box.

No. DK1144 Soldering Outfit Size 12¾x1½x2 .... **$0.40**

Dear Sirs:—   Easton, Pa.

I am now in receipt of the Solderall. I have tried it and it works fine, and am well pleased with it.   **WILSON PAULUS.**

The "Electro" Toy Transformer
Step Down Transformer for Alternating Current Only
CAPACITY 50 WATTS

This transformer is designed to permit the operation of toy motors, railways, lamps, etc., and to do so continuously under the severest service a device of this kind can undergo. The small boy is very prone to give an article of this kind very hard bumps and short circuits and yet we know our transformer will stand up and work under it all. It has a very heavy controller handle and the only contacts exposed are the low voltage binding posts, controller contacts and 110 volt connections. The transformer gives 3 voltages as follows: 6, 9 or 12 volts, by simply moving a lever, and motor speeds can be controlled in this manner.

It will furnish 4 amperes at any one of the three voltages or a maximum capacity of 48 watts.

No. EFE6707

The finish is black enamel and is very attractive. Transformer is complete with 6 ft. of flexible cord and an attachment plug. Built for operation on 110-125 volt circuits and 60 cycles. Size 5¼x3x4¼ in.

No. EFE6707 "Electro" Toy Transformer, 50 watt capacity, as described **$5.65**

Shipping weight 8 lbs.

The "Electro" Low Voltage Transformers
Step Down Transformer for Alternating Current Only
CAPACITY 32 WATTS

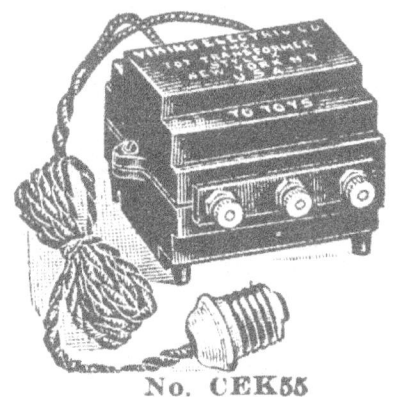

This is a new type of low voltage alternating current transformer which has been evolved by us and is no doubt, to-day, the most efficient as well as cheapest on the market. This transformer can only be used on alternating current up to 110 volts, 60 cycle. It reduces the 110 volt current to 2-6 or 8 volts and gives 4 amperes at any one of the voltages. Spring contact connections are provided, so that these voltages can be taken off. This transformer can be used to operate small railways, small motors, bells, spark coils, electro magnets, telegraph instruments, lamps, etc., etc. It is understood, however, that storage batteries can not be recharged with this transformer as it naturally only furnishes alternating current. For the uses enumerated above, however, this transformer will do wonders and will replace storage cells and dry cells in a great many instances. The cost of operation is almost negligible as the transformer uses much less current than a 16 c.p. lamp. The output of the transformer is 32 watts.

No. CEK55

The apparatus is enclosed in a steel case, and the windings are immersed in an insulating compound. Each transformer is equipped with plug and 8 feet of flexible cord, so that it may be attached to any alternating current lamp socket. It has no moving parts to get out of order and, with ordinary care, will last indefinitely. The size is 4¼x4½x3¼ in.

No. CEK55 Transformer, as described, complete **$3.50**

Shipping weight 8 lbs.

Gentlemen:— Louisville, Ohio.
The goods which I purchased from you ARRIVED ALL O. K. The Spark Coil DESERVES PRAISE. I intend sending you an order in the near future. Will you please send me a catalogue, giving prices on all wireless goods.   Respectfully,   SANFORD ESSIG.

Carbon Diaphragms, Back Plates, etc.

The Diaphragms we list are standard and are 1/64 inches in thickness. Diameter invariably is 2⅛ in. These diaphragms are used in all kinds of sensitive transmitter work and are guaranteed to be of the best material.

No. BK6084   Carbon Diaphragms, as described, each............. **$0.20**

No. AK6083   Iron Telephone Diaphragms, 2⅛ in. diameter, as used in our low-priced receivers, each **$0.10**

No. CE6090   Carbon "Back Plate" with 6 cup-shaped indentations to take our globular carbon, has hole in center for screw to pass, each **$0.35**

Shipping weight any of above 4 oz.

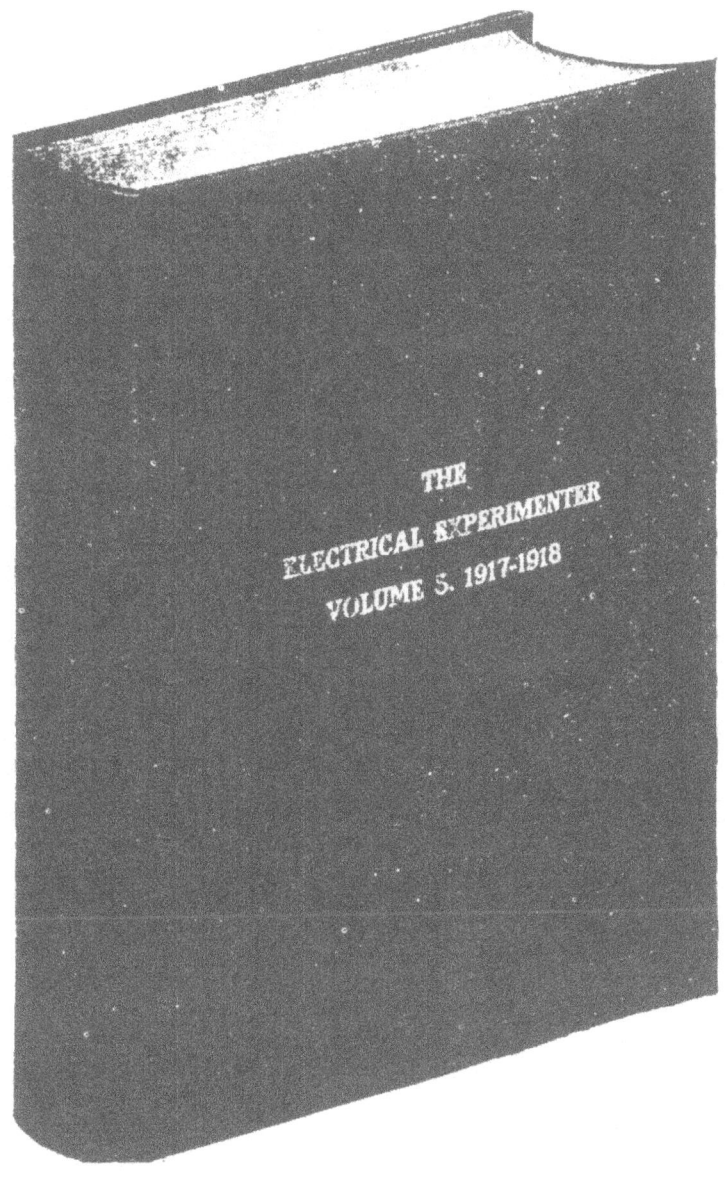

Bound Volume of "The Electrical Experimenter"

Science and Invention

This book cost $28,000. Over 1,200 authors contributed.

Contains a goldmine of electrical and scientific information. No such value has ever been offered before for so low a price. A marvelous cyclopedia of electricity. A reference book of authentic information not found in any other book in print.

Volume contains twelve numbers of the "Electrical Experimenter" magazine—May, 1917, to April, 1918, inclusive. There are 928 pages, 2,110 complete articles, 1,908 illustrations, 369 questions and answers. Size, 12" high; 9" wide; 1¾" thick.

Bound Volume as described ... **$2.00**

Shipping weight 8 lbs.

HAWKINS ELECTRICAL GUIDES.

POCKET SIZE.

$1.00 Each Postpaid.

Here is a set of books that nobody interested in Electricity should do without.

Hawkins Electrical Guides are a school within themselves, containing a complete study course, with **Questions, Answers** and **Illustrations** written in plain everyday language so that you can understand the "How, WHEN AND WHY" of ELECTRICITY.

They are handsomely bound in flexible black Buckram, with gold edges, and will readily go in the pocket.

Each book is complete in itself and will be supplied at $1 per copy.

Read over the following titles and note the scope of each book:

HAWKINS' ELECTRICAL GUIDE NO. 1.

Contains 240 pages, 246 illustrations and explains fully by Questions, Answers and Illustrations all about Electrical Signs and Symbols; Static and Current Electricity; Primary Cells; Conductors and Insulators; Resistance and Conductivity; Effects of Current; Magnetism; Electro-Magnetic Induction; Induction Coils; Dynamo Principles; Different Classes of Dynamos; Field Magnets.

HAWKINS' ELECTRICAL GUIDE NO. 2.

Contains 218 pages, 256 pictures and explains in very easily understood manner what you should know about Armature Principles; Armature Winding; The Theory of Armatures; Commutators and Commutation; Brushes and the Brush Gear; Armature Construction; Motor Principles; Armature Reaction in Motors; Starting a Motor; Motor Calculations; Brake Horse Power; Selection and Installations of Motors and Dynamos; Performance Curves; Auxiliary Apparatus.

HAWKINS' ELECTRICAL GUIDE NO. 3.

Contains 276 pages, 280 illustrations and complete instruction on the following subjects: Galvanometers; Standard Cells; Current Measurement; Resistance Measurement; Christie Bridge; Testing Sets; Loop Tests; Potentiometer; Armature Voltmeter and Wattmeter; Multipliers; Electro-Dynamometers; Demand Indicators; Watt-Hour Meters; Operation of Dynamos; Lubrication; Troubles; Coupling of Dynamos; Armature Troubles; Care of Commutator and Brushes; Heating; Operating of Motors; Starters; Speed Regulators.

HAWKINS' ELECTRICAL GUIDE NO. 4.

Contains 308 pages, 423 illustrations and gives very useful information on the following important subjects: Distribution Systems; Boosters; Wires and Wire Calculations; Inside, Outside and Underground Wiring; Wiring of Buildings; Sign Flashers; Lightning Protection; Storage Battery; Rectifiers; Storage Battery Systems.

HAWKINS' ELECTRICAL GUIDE NO. 5.

Contains 320 pages, 614 illustrations. Alternating Current Motors; Synchronous and Induction Motor Principles; A.C. Commutator Motors; Induction Motors; Transformers, Losses, Construction, Connections, Tests; Converters, Rotary, Voltage Regulation, Frequency Changing Sets, Parallel Operation, Cascade Converters; Rectifiers, Mechanical, Electrolytic, Electromagnetic; Alternating Current Systems.

HAWKINS' ELECTRICAL GUIDE NO. 6.

Contains 298 pages, 472 illustrations. Alternating Current Systems; Switching Devices; Circuit Breakers; Relays; Lightning Protection Apparatus; Regulating Devices; Synchronous Condensers; Indicating Devices; Meters; Power Factor Indicators; Wave Form Measurement; Switchboards. etc.

HAWKINS' ELECTRICAL GUIDE NO. 7.

Contains 316 pages. 379 illustrations. Alternating Current Wiring; A. C. Wiring Calculations; Table: Properties of Copper Wire; Power Stations; Hydro-Electric Plants; Isolated Plants; Sub-Station Management; Turbines; Selection, Location, Erection, Running, Care and Repair; Station Testing; Telephones; Principles and Construction; Various Systems; Wiring Diagrams; Telephone Troubles.

HAWKINS' ELECTRICAL GUIDE NO. 8.

Contains 332 pages, 436 illustrations. Telegraph; Simultaneous Telegraphy and Telephony; Wireless Principles, Construction, Diagrams; Electric Bells; Electric Lighting; Illumination; Photometry, etc.

HAWKINS' ELECTRICAL GUIDE NO. 9.

Contains 576 pages, 849 illustrations and treating thoroughly on the following important subjects: Telephones; Telegraph; Simultaneous Telegraphy and Telephony; Wireless; Electric Bells; Electric Lighting; Photometry; Electric Railways; Electric Locomotives; Car Lighting; Trolley Car Operation.

HAWKINS' ELECTRICAL GUIDE NO. 10.

Contains 704 pages, 730 illustrations, is the last number of the series and completes this very remarkable series, including the 125 pages of ready reference index of the 10 numbers. The subjects that are covered are: Miscellaneous Applications; Motion Pictures; Gas Engine Ignition; Automobile Self-Starters and Lighting Systems; Electric Vehicles; Elevators; Cranes; Pumps; Air Compressors; Electric Heating; Electric Welding; Soldering and Brazing; Industrial Electrolysis; Electro-Plating; Electro-Therapeutics, X Rays, etc.

This number contains A COMPLETE READY REFERENCE INDEX OF THE COMPLETE LIBRARY.

Price $1.00 a Book Prepaid

IMMEDIATE SHIPMENTS

WHERE ORDERS ARE RECEIVED AND RECORDED.

ENORMOUS STOCKS MAKE PROMPT SHIPMENTS POSSIBLE.

ASSEMBLERS' DEPARTMENT IN FACTORY.

WHERE YOUR ORDER IS FILLED WITHIN 12 HOURS.

ORDER BLANK

To the **ELECTRO IMPORTING CO.**

233 Fulton Street

New York City

One Lesson of our famous WIRE-LESS COURSE is given for Each Dollar's worth of goods you buy. Make your order for at least $1.00 so you can obtain this wonderful premium.

(See information in yellow section of this catalogue.)

Please send the following to:

NAME .... BOX ....

POSTOFFICE .... R. F. D. ....

STREET ADDRESS ....

COUNTY .... STATE ....

Express Office if Different from Postoffice.

Fill This In ☞	How do you prefer your goods to come? Mark X in proper place.		EXPRESS What Company?			
	INSURED MAIL				Dollars	Cents
PARCEL POST At customer's risk of loss or damage.	Guarantees s a f e delivery. Add 3c. for insurance.		How much are you enclosing with this order?			
If your order is over $3.00 we will at your request ship it to you C. O. D., but to show your good will, you are required to make a deposit of 25% of the amount.	Express C. O. D. shipments may be inspected at arrival. Postal C. O. D. shipments are not allowed to be examined.			C. O. D.		

Order Number of Article in Catalogue	Quantity Desired	NAME OF ARTICLE WANTED	PRICE		TOTAL	

ORDER BLANK (Continued)

Order Number of Article in Catalogue	Quantity Desired	NAME OF ARTICLE WANTED	PRICE	TOTAL
		Brought forward		
		Postage		
		Insurance		
Positively no order for less than 50 Cents accepted.		Total	$	$

BEFORE ORDERING READ THIS:

☞ NO ORDER FOR LESS THAN 50 CENTS ACCEPTED ☜

SAFEST WAY TO SEND MONEY. All orders must be accompanied by money to avoid delay to you. The best way to send money: Post Office Money Order, Express Money Order, Bank Draft (ON NEW YORK BANKS), Cash by Registered Mail. Money sent in any other way is AT YOUR RISK. Cash or Stamps must be sent registered, as we are not responsible for any loss if sent unregistered.

We accept stamps (if new) in ANY quantity, but in amounts over $3.00 you must include 5 per cent. extra, which amount we must pay to stamp brokers.

Mutilated or ungummed stamps, likewise Canadian or other foreign stamps, will be promptly RETURNED TO YOU. Never send Canadian or other foreign bills or coins, or slick or mutilated coins. It surely delays your order, as we INVARIABLY return money of this kind. Coins should be wrapped carefully, else they will break through the envelope. STAMPS SHOULD BE PLACED BETWEEN SHEETS OF WAX OR PARAFFINE PAPER.

We prefer to receive the money in Post Office or Express Money Orders. This only costs you a few cents; besides, YOU GET A RECEIPT. And if the money order should be lost or destroyed, you get a new one (duplicate) at no cost to you.

PERSONAL CHECKS are not accepted, unless you add 10 cents, which amount we must pay to our bank to collect the money. If the check is not certified by your bank, we DO NOT SHIP the order until we receive advice from our bank that the check has been cashed. This means delay; consequently, if you desire prompt shipment, HAVE YOUR CHECK CERTIFIED. It costs you nothing to do so.

INSURANCE ON PARCELS.

Fragile articles will be carefully packed and duly labeled by us, but as the Parcel Post does not guarantee their safe delivery, we cannot be held responsible for breakage or lost shipments. For your own protection, order Parcel Post goods INSURED. The fee for this insurance for each package is:

$0.03 for $10.— Insurance.
$0.05 for $25.— Insurance.
$0.10 for $50.— Insurance.

Always allow sufficient money to cover postage. Weight of packages can be easily figured from the weights given in the Catalogue descriptions. Knowing the weight and the parcel post zone in which your postoffice is located, measuring from New York, you can easily figure the amount of postage required from the parcel post rate table shown in the front section of this catalogue.

THREE WONDERFUL BOOKS

How to Make Wireless Sending Apparatus

Contains information on how to make 30 different pieces of wireless sending apparatus from materials easily obtained. Illustrations and descriptions are big, simple, and easily understood.

Only modern apparatus is described by 20 wireless experts who give you the benefit of their experience.

Tells How-to-Make an Experimental Arc Set, Speaking Arc, Quenched Gap, ¼ K.W. Transformer, Oscillation Transformer, Photophone, etc., etc.

Book has 100 pages, (size 7x5 in.) 88 illustrations, paper cover printed in two colors.

No. BE140

No. BE140   How to Make Wireless Sending Apparatus. Price prepaid...... **$0.25**

How to Make Wireless Receiving Apparatus

Written entirely for the Wireless "Bug" who makes his own apparatus. The 20 radio constructors who wrote the book know how articles should be made from simple materials.

Only modern apparatus is described such as Receiving Set without aerial or ground, magnetic detector, wireless relay, wireless lecture set, etc., etc. Book has 100 pages, each 5x7 inches, 90 illustrations, many full pages, paper bound in two colors. A wonderful book and one you need.

No. BE141   How to Make Wireless Receiving Apparatus. Price prepaid...... **$0.25**

No. BE141.

Experimental Electricity Course

A masterpiece. Just the book you need to tell you all about electricity and electrical facts in plain everyday language that you can understand. Explains every electrical device from a push button and bell to the biggest generator made. Worth its weight in gold for the man who doesn't know enough about electricity, and to the experimenter it is still more valuable because of its many facts, tables, etc., etc.

Book has a stiff cloth cover, is 5x9 inches in size, and contains 160 pages, 400 illustrations and is so shaped that it just slips in your pocket to read while you ride. See back cover of this catalog for a wonderful offer.

A real bargain at the price.

No. AX103

No. AX103 Experimental Electricity Course, sent prepaid......... **$1.00**

Keeps You Up-to-Date on Electricity, Science, Wireless.

THE ELECTRICAL EXPERIMENTER SCIENCE AND INVENTION

Devoted solely to the interests of the electrical experimenter, prints nothing but articles on Wireless and Electricity. It is clean, up-to-date and original. The magazine you must have.

Contains new articles on Wireless and Electricity every issue, also the following departments: "The Constructor," "Wireless Department," "How to Make It," "Latest Patents," "Phoney Patents," "Among the Amateurs," "Question Box," "Patent Advice," "Experimental Chemistry," "Marvels of Physics" and good scientific fiction. Every article is by an authority.

Every issue contains at least 84 big 9 x 12 inch pages (⅞ as large as Saturday Evening Post) and 150 to 200 original illustrations that are only possible on so large a page. The magazine is edited by H. Gernsback.

PUBLISHED MONTHLY—12 NUMBERS A YEAR

PRICE $2.00 A YEAR

Send postal for a sample copy

New York City, Canada and Foreign Countries 50c per year extra

SPECIAL OFFER

As a special offer we will give you free for a limited time only, EITHER a complete copy of the 160 page 20 lesson *"WIRELESS COURSE" or a cloth bound copy of the †"EXPERIMENTAL ELECTRICITY COURSE" with a year's subscription (12 numbers) of the Electrical Experimenter at $2.00. Either of these books is worth more than the subscription price, yet you will get 12 big numbers of this wonderful magazine and either big book FREE for only $2.00. Don't delay. This big offer can't hold good long; better send your order to-day and we will mail you the book FREE and prepaid by return mail.

*For description of Wireless Course see page 100. †For description of Experimental Electricity Course see inside back cover of this catalog.

FOR SALE ON ALL NEWS-STANDS, 20c PER COPY

Address all communications, contributions, etc., to "The Electrical Experimenter."
Money orders, checks (if out-of-town checks add 10 cents for exchange) must be drawn to the order of Experimenter Publishing Co., Inc.

Experimenter Publishing Co., Inc.   233 W. Fulton Street New York City

www.ingramcontent.com/pod-product-compliance
Lightning Source LLC
Chambersburg PA
CBHW080808180526
45168CB00006B/2373